PRAISE FOR THE *MANGA GUIDE* SERIES

"Highly recommended."
—CHOICE MAGAZINE ON *THE MANGA GUIDE TO DATA*

"Stimulus for the next generation of scientists."
—SCIENTIFIC COMPUTING ON *THE MANGA GUIDE TO MOLECULAR BIOLOGY*

"A great fit of form and subject. Recommended."
—OTAKU USA MAGAZINE ON *THE MANGA GUIDE TO PHYSICS*

"The art is charming and the humor engaging. A fun and fairly painless lesson on what many consider to be a less-than-thrilling subject."
—SCHOOL LIBRARY JOURNAL ON *THE MANGA GUIDE TO STATISTICS*

"This is really what a good math text should be like. Unlike the majority of books on subjects like statistics, it doesn't just present the material as a dry series of pointless-seeming formulas. It presents statistics as something *fun*, and something enlightening."
—GOOD MATH, BAD MATH ON *THE MANGA GUIDE TO STATISTICS*

"I found the cartoon approach of this book so compelling and its story so endearing that I recommend that every teacher of introductory physics, in both high school and college, consider using it."
—AMERICAN JOURNAL OF PHYSICS ON *THE MANGA GUIDE TO PHYSICS*

"The series is consistently good. A great way to introduce kids to the wonder and vastness of the cosmos."
—DISCOVERY.COM ON *THE MANGA GUIDE TO THE UNIVERSE*

"A single tortured cry will escape the lips of every thirty-something biochem major who sees *The Manga Guide to Molecular Biology*: 'Why, oh why couldn't this have been written when I was in college?'"
—THE SAN FRANCISCO EXAMINER

"Scientifically solid . . . entertainingly bizarre."
—CHAD ORZEL, AUTHOR OF *HOW TO TEACH PHYSICS TO YOUR DOG*, ON *THE MANGA GUIDE TO RELATIVITY*

"A lot of fun to read. The interactions between the characters are lighthearted, and the whole setting has a sort of quirkiness about it that makes you keep reading just for the joy of it."
—HACK A DAY ON *THE MANGA GUIDE TO ELECTRICITY*

"*The Manga Guide to Databases* was the most enjoyable tech book I've ever read."
—RIKKI KITE, LINUX PRO MAGAZINE

"The *Manga Guides* definitely have a place on my bookshelf."
—SMITHSONIAN'S "SURPRISING SCIENCE"

"For parents trying to give their kids an edge or just for kids with a curiosity about their electronics, *The Manga Guide to Electricity* should definitely be on their bookshelves."
—SACRAMENTO BOOK REVIEW

"This is a solid book and I wish there were more like it in the IT world."
—SLASHDOT ON *THE MANGA GUIDE TO DATABASES*

"*The Manga Guide to Electricity* makes accessible a very intimidating subject, letting the reader have fun while still delivering the goods."
—GEEKDAD BLOG, WIRED.COM

"If you want to introduce a subject that kids wouldn't normally be very interested in, give it an amusing storyline and wrap it in cartoons."
—MAKE ON *THE MANGA GUIDE TO STATISTICS*

"A clever blend that makes relativity easier to think about—even if you're no Einstein."
—STARDATE, UNIVERSITY OF TEXAS, ON *THE MANGA GUIDE TO RELATIVITY*

"This book does exactly what it is supposed to: offer a fun, interesting way to learn calculus concepts that would otherwise be extremely bland to memorize."
—DAILY TECH ON *THE MANGA GUIDE TO CALCULUS*

"The art is fantastic, and the teaching method is both fun and educational."
—ACTIVE ANIME ON *THE MANGA GUIDE TO PHYSICS*

"An awfully fun, highly educational read."
—FRAZZLEDDAD ON *THE MANGA GUIDE TO PHYSICS*

"Makes it possible for a 10-year-old to develop a decent working knowledge of a subject that sends most college students running for the hills."
—SKEPTICBLOG ON *THE MANGA GUIDE TO MOLECULAR BIOLOGY*

"This book is by far the best book I have read on the subject. I think this book absolutely rocks and recommend it to anyone working with or just interested in databases."
—GEEK AT LARGE ON *THE MANGA GUIDE TO DATABASES*

"The book purposefully departs from a traditional physics textbook and it does it very well."
—DR. MARINA MILNER-BOLOTIN, RYERSON UNIVERSITY ON *THE MANGA GUIDE TO PHYSICS*

"Kids would be, I think, much more likely to actually pick this up and find out if they are interested in statistics as opposed to a regular textbook."
—GEEK BOOK ON *THE MANGA GUIDE TO STATISTICS*

THE MANGA GUIDE™ TO LINEAR ALGEBRA

THE MANGA GUIDE™ TO LINEAR ALGEBRA

SHIN TAKAHASHI,
IROHA INOUE, AND
TREND-PRO CO., LTD.

The Manga Guide to Linear Algebra is a translation of the Japanese original, *Manga de wakaru senkeidaisuu*, published by Ohmsha, Ltd. of Tokyo, Japan, © 2008 by Shin Takahashi and TREND-PRO Co., Ltd.

This English edition is co-published by No Starch Press, Inc. and Ohmsha, Ltd.

First printing

16 15 14 13 12 1 2 3 4 5 6 7 8 9

ISBN-10: 1-59327-413-0
ISBN-13: 978-1-59327-413-9

Publisher: William Pollock
Author: Shin Takahashi
Illustrator: Iroha Inoue
Producer: TREND-PRO Co., Ltd.
Production Editor: Alison Law
Developmental Editor: Keith Fancher
Translator: Fredrik Lindh
Technical Reviewer: Eric Gossett
Compositor: Riley Hoffman
Proofreader: Paula L. Fleming
Indexer: BIM Indexing & Proofreading Services

For information on book distributors or translations, please contact No Starch Press, Inc. directly:
No Starch Press, Inc.
38 Ringold Street, San Francisco, CA 94103
phone: 415.863.9900; fax: 415.863.9950; info@nostarch.com; http://www.nostarch.com/

Library of Congress Cataloging-in-Publication Data
Takahashi, Shin.
[Manga de wakaru senkei daisu. English]
The manga guide to linear algebra / Shin Takahashi, Iroha Inoue, Trend-pro Co. Ltd.
p. cm.
ISBN 978-1-59327-413-9 (pbk.) -- ISBN 1-59327-413-0 (pbk.)
1. Algebras, Linear--Comic books, strips, etc. 2. Graphic novels. I. Inoue, Iroha. II. Trend-pro Co. III. Title.
QA184.2.T3513 2012
512'.50222--dc23

2012012824

CONTENTS

PREFACE

This book is for anyone who would like to get a good overview of linear algebra in a relatively short amount of time.

Those who will get the most out of *The Manga Guide to Linear Algebra* are:

- University students about to take linear algebra, or those who are already taking the course and need a helping hand
- Students who have taken linear algebra in the past but still don't really understand what it's all about
- High school students who are aiming to enter a technical university
- Anyone else with a sense of humor and an interest in mathematics!

The book contains the following parts:

Chapter 1: What Is Linear Algebra?
Chapter 2: The Fundamentals
Chapters 3 and 4: Matrices
Chapters 5 and 6: Vectors
Chapter 7: Linear Transformations
Chapter 8: Eigenvalues and Eigenvectors

Most chapters are made up of a manga section and a text section. While skipping the text parts and reading only the manga will give you a quick overview of each subject, I recommend that you read both parts and then review each subject in more detail for maximal effect. This book is meant as a complement to other, more comprehensive literature, not as a substitute.

I would like to thank my publisher, Ohmsha, for giving me the opportunity to write this book, as well as Iroha Inoue, the book's illustrator. I would also like to express my gratitude towards re_akino, who created the scenario, and everyone at Trend Pro who made it possible for me to convert my manuscript into this manga. I also received plenty of good advice from Kazuyuki Hiraoka and Shizuka Hori. I thank you all.

SHIN TAKAHASHI
NOVEMBER 2008

PROLOGUE
LET THE TRAINING BEGIN!

HANAMICHI UNIVERSITY

エイッ

EI!!

セヤァァ

SEYAAA!

OKAY!

真派花道空手会*

IT'S NOW OR NEVER!

NOTHING TO BE AFRAID OF...

* HANAMICHI KARATE CLUB

ドキ
ドキ
BA-DUM
BA-DUM
RATTLE
EXCU–
BUMP
WHA–
真派花道

WHAT DO WE HAVE HERE?

I... I'M A FRESHMAN... MY NAME IS REIJI YURINO.
WOULD YOU BY ANY CHANCE BE TETSUO ICHINOSE, THE KARATE CLUB CAPTAIN?
INDEED.
WOW... THE HANAMICHI HAMMER HIMSELF!
U-UM...
?
I CAN'T BACK DOWN NOW!
I WANT TO JOIN THE KARATE CLUB!
BOW
I DON'T HAVE ANY EXPERIENCE, BUT I THINK I—
YANK
ARE YOU SERIOUS? MY STUDENTS WOULD CHEW YOU UP AND SPIT YOU OUT.

PLEASE! I—
I WANT TO GET STRONGER!
SHAKE
SHAKE
...
HMM?
HAVEN'T I SEEN YOUR FACE SOMEWHERE?
UHH...
AHA!

AREN'T YOU THAT GUY? THE ONE ON MY SISTER'S MATH BOOK?
現役大学生が教える
誰でも徹底数学
百合野玲治著
MATHEMATICS FOR EVERYONE
AUTHOR: REIJI YURINO
BY STUDENTS—FOR STUDENTS
OH, YOU'VE SEEN MY BOOK?

SO IT IS YOU!
Y-YES.
I MAY NOT BE THE STRONGEST GUY...
BUT I'VE ALWAYS BEEN A WHIZ WITH NUMBERS.
I SEE...
HMM
I MIGHT CONSIDER LETTING YOU INTO THE CLUB...
WHA—?

REALLY?!
...UNDER ONE CONDITION!

YOU HAVE TO TUTOR MY LITTLE SISTER IN MATH.
ERR
SHE'S NEVER BEEN THAT GOOD WITH NUMBERS, YOU SEE...
AND SHE COMPLAINED JUST YESTERDAY THAT SHE'S BEEN HAVING TROUBLE IN HER LINEAR ALGEBRA CLASS...
SO IF I TUTOR YOUR SISTER YOU'LL LET ME IN THE CLUB?

真派花道
真派
WOULD THAT BE ACCEPTABLE TO YOU?
OF COURSE!
I SUPPOSE I SHOULD GIVE YOU FAIR WARNING...

IF YOU TRY TO MAKE A PASS AT HER...
EVEN ONCE...
CRACK
SNAP
ゴキ
ゴキ
I... WOULDN'T THINK OF IT!

IN THAT CASE... FOLLOW ME.
WE WON'T GO EASY ON YOU, YA KNOW.
OF COURSE!
WE'LL START RIGHT AWAY!
真派花道空手会
I'M IN!

1
WHAT IS LINEAR ALGEBRA?

* *OSSU* IS AN INTERJECTION OFTEN USED IN JAPANESE MARTIAL ARTS TO ENHANCE CONCENTRATION AND INCREASE THE POWER OF ONE'S BLOWS.

ストーー
MMH...
THUMP
I WONDER WHAT SHE'LL BE LIKE...
OWW.
SHE LOOKS A LITTLE LIKE TETSUO...
UMM...
EXCUSE ME, ARE YOU REIJI YURINO?
WHA—
PLEASED TO MEET YOU.
I'M MISA ICHINOSE.

SORRY ABOUT MY BROTHER ASKING YOU TO DO THIS ALL OF A SUDDEN.
NO PROBLEM! I DON'T MIND AT ALL!
HOW COULD THIS GIRL POSSIBLY BE HIS SISTER?!
BUT...
I HAD NO IDEA THAT I WENT TO SCHOOL WITH THE FAMOUS REIJI YURINO!

WELL, I WOULDN'T EXACTLY SAY *FAMOUS*...
I'VE BEEN LOOKING FORWARD TO THIS A LOT!
ERR... WOULD IT BE AWKWARD IF I ASKED YOU TO SIGN THIS?

LIKE...AN AUTOGRAPH?
ONLY IF YOU WANT TO. IF IT'S TOO WEIRD—
NO, IT'S MY PLEASURE.
TOUCH
GULP.

SO–
...
SOR–
–RY
SORRY MISA, BUT I THINK I'LL HEAD HOME A BIT EARLY TODAY.
BIG BROTHER!
YURINOOO
YOU REMEMBER OUR LITTLE TALK?
YEP!

AN OVERVIEW OF LINEAR ALGEBRA

THE CALCULATIONS AREN'T NEARLY AS HARD AS THEY LOOK!
AND ONCE YOU UNDERSTAND THE BASICS, THE MATH BEHIND IT IS ACTUALLY VERY SIMPLE.
REALLY?

I WOULDN'T SAY IT'S MIDDLE SCHOOL LEVEL, BUT IT'S NOT FAR OFF.

OH! WELL, THAT'S A RELIEF...
YOU SAID IT!

BUT I STILL DON'T UNDERSTAND... WHAT *IS* LINEAR ALGEBRA EXACTLY?
UM...

ERR...

THAT'S A TOUGH QUESTION TO ANSWER PROPERLY.
REALLY? WHY?
WELL, IT'S PRETTY ABSTRACT STUFF. BUT I'LL GIVE IT MY BEST SHOT.

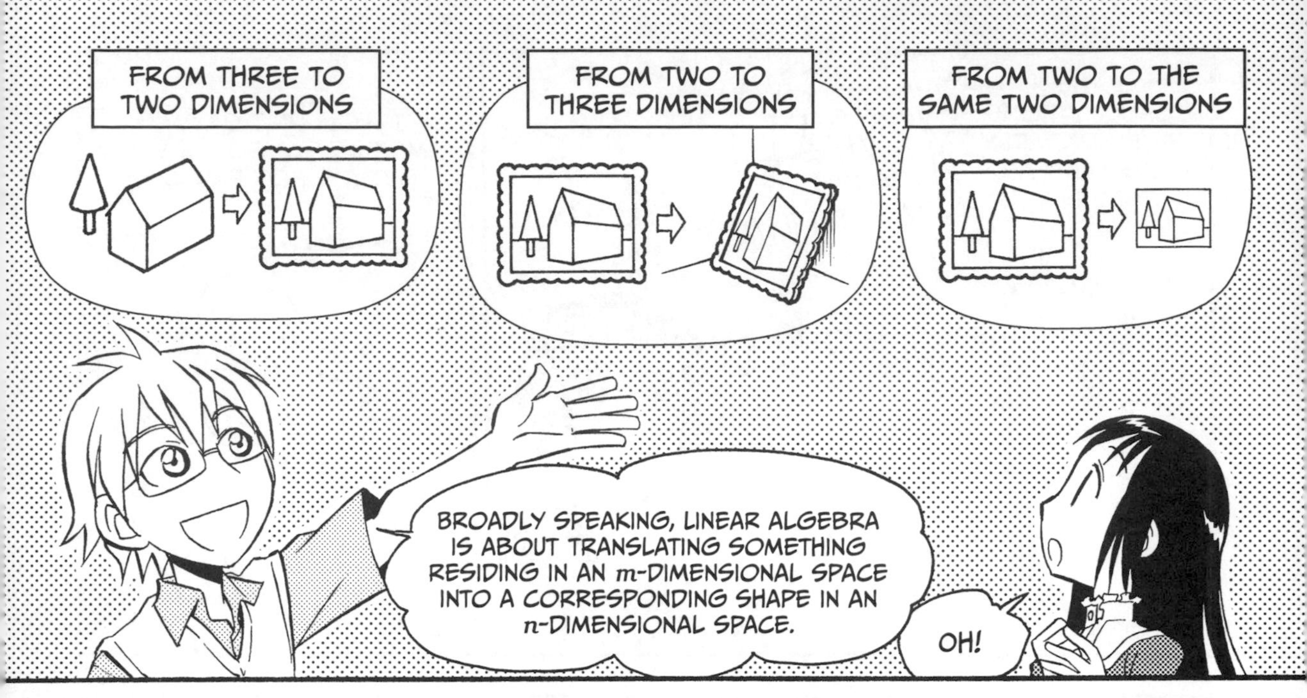
FROM THREE TO TWO DIMENSIONS
FROM TWO TO THREE DIMENSIONS
FROM TWO TO THE SAME TWO DIMENSIONS
BROADLY SPEAKING, LINEAR ALGEBRA IS ABOUT TRANSLATING SOMETHING RESIDING IN AN m-DIMENSIONAL SPACE INTO A CORRESPONDING SHAPE IN AN n-DIMENSIONAL SPACE.
OH!

WE'LL LEARN TO WORK WITH MATRICES...
MATRICES
VECTORS
AND VECTORS...

WITH THE GOAL OF UNDERSTANDING THE CENTRAL CONCEPTS OF:
· LINEAR TRANSFORMATIONS
· EIGENVALUES AND EIGENVECTORS
LINEAR TRANSFORMATIONS
EIGENVALUES AND EIGENVECTORS
VECTORS
MATRICES
I SEE...

SO...
?

WHAT EXACTLY IS IT GOOD FOR? OUTSIDE OF ACADEMIC INTEREST, OF COURSE.
...

YOU JUST HAD TO ASK ME THE DREADED QUESTION, DIDN'T YOU?
EH?

WHILE IT IS USEFUL FOR A MULTITUDE OF PURPOSES INDIRECTLY, SUCH AS EARTHQUAKE-PROOFING ARCHITECTURE, FIGHTING DISEASES, PROTECTING MARINE WILDLIFE, AND GENERATING COMPUTER GRAPHICS...
IT DOESN'T STAND THAT WELL ON ITS OWN, TO BE COMPLETELY HONEST.
OH?

AND MATHEMATICIANS AND PHYSICISTS ARE THE ONLY ONES WHO ARE REALLY ABLE TO USE THE SUBJECT TO ITS FULLEST POTENTIAL.
AWW!

SO EVEN IF I DECIDE TO STUDY, IT WON'T DO ME ANY GOOD IN THE END?
THAT'S NOT WHAT I MEANT AT ALL!

FOR EXAMPLE, FOR AN ASPIRING CHEF TO EXCEL AT HIS JOB, HE HAS TO KNOW HOW TO FILLET A FISH; IT'S JUST CONSIDERED COMMON KNOWLEDGE.
THE SAME RELATIONSHIP HOLDS FOR MATH AND SCIENCE STUDENTS AND LINEAR ALGEBRA; WE SHOULD ALL KNOW HOW TO DO IT.
I SEE...

LIKE IT OR NOT, IT'S JUST ONE OF THOSE THINGS YOU'VE GOT TO KNOW.
BEST NOT TO FIGHT IT. JUST BUCKLE DOWN AND STUDY, AND YOU'LL DO FINE.
THERE IS ALSO A LOT OF ACADEMIC LITERATURE THAT YOU WON'T UNDERSTAND IF YOU DON'T KNOW LINEAR ALGEBRA.
I'LL TRY!

REGARDING OUR FUTURE LESSONS...
I THINK IT WOULD BE BEST IF WE CONCENTRATED ON UNDERSTANDING LINEAR ALGEBRA AS A WHOLE.
LINEAR TRANSFORMATIONS
EIGENVALUES AND EIGENVECTORS
VECTORS
MATRICES
わー

MOST BOOKS AND COURSES IN THE SUBJECT DEAL WITH LONG CALCULATIONS AND DETAILED PROOFS.
WHICH LEADS TO...
I'LL TRY TO AVOID THAT AS MUCH AS POSSIBLE...
WHEW.

...AND FOCUS ON EXPLAINING THE BASICS AS BEST I CAN.
GREAT!

WELL, I DON'T WANT TO OVERWHELM YOU. WHY DON'T WE CALL IT A DAY?
WE'LL START WITH SOME OF THE FUNDAMENTALS NEXT TIME WE MEET.
OKAY.
RING
RING
OOPS, THAT'S ME!
RING
RING
IT'S FROM MY BROTHER.
OH?
THE SENSEI TEXTS?!
HARD TO IMAGINE...
"COME HOME NOW"
WELL, UH... TELL HIM I SAID HELLO!

2
THE FUNDAMENTALS

97
98
99
YOU'VE GOT TO KNOW YOUR BASICS!
GGHH
WHUMP
D-DONE!
100...
YOU WISH! AFTER YOU'RE DONE WITH THE PUSHUPS, I WANT YOU TO START ON YOUR LEGS! THAT MEANS SQUATS! GO GO GO!

HEY... I THOUGHT WE'D START OFF WITH...
REIJI, YOU SEEM PRETTY OUT OF IT TODAY.
ARE YOU OKAY?
I-I'LL BE FINE. TAKE A LOOK AT THI—

RRRRUMBLE
AH—

SORRY, I GUESS I COULD USE A SNACK...
DON'T WORRY, PUSHING YOUR BODY THAT HARD HAS ITS CONSEQUENCES.

JUST GIVE ME FIVE MINUTES...
I DON'T MIND. TAKE YOUR TIME.

OM NOM NOM

WELL THEN, LET'S BEGIN.
TAKE A LOOK AT THIS.

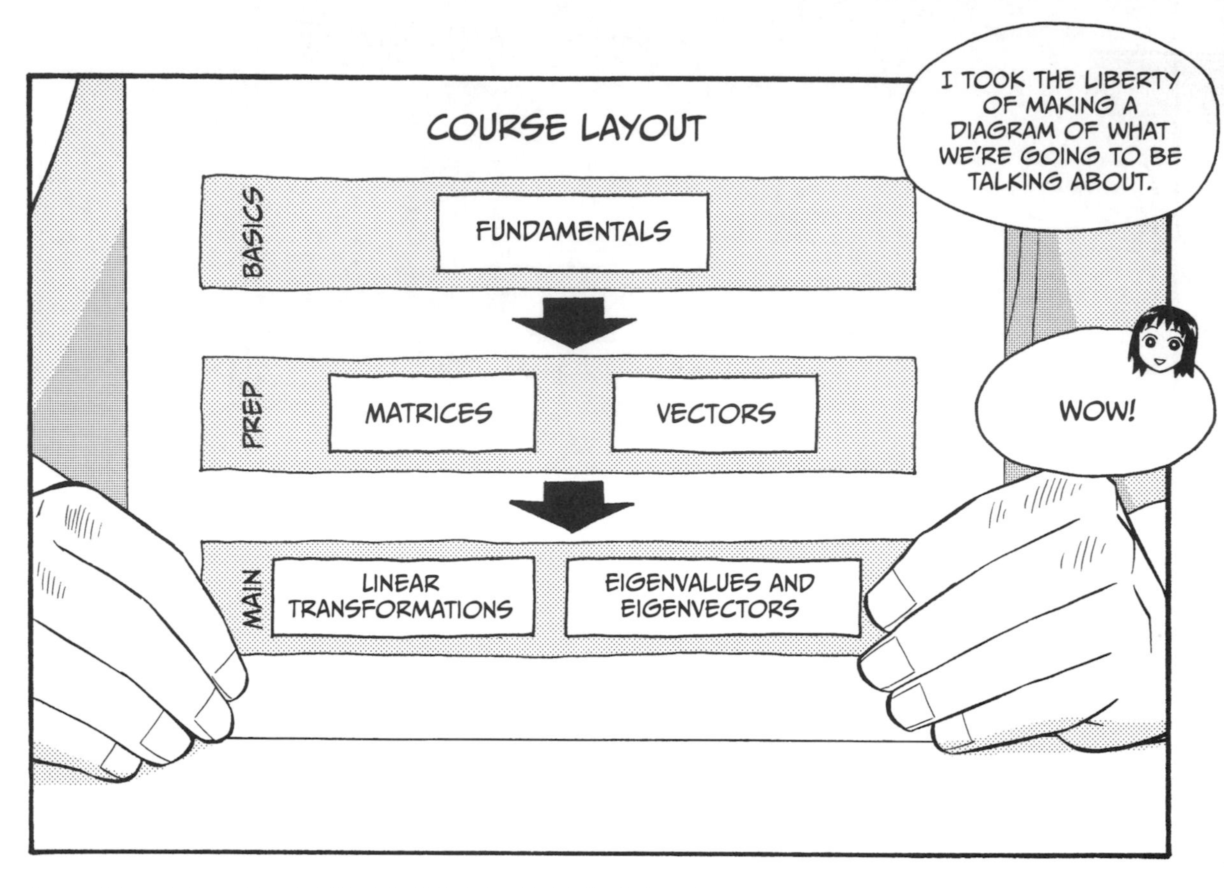
COURSE LAYOUT
BASICS
FUNDAMENTALS
PREP
MATRICES
VECTORS
MAIN
LINEAR TRANSFORMATIONS
EIGENVALUES AND EIGENVECTORS
I TOOK THE LIBERTY OF MAKING A DIAGRAM OF WHAT WE'RE GOING TO BE TALKING ABOUT.
WOW!

I THOUGHT TODAY WE'D START ON ALL THE BASIC MATHEMATICS NEEDED TO UNDERSTAND LINEAR ALGEBRA.
COURSE LAYOUT
FUNDAMENTALS
MATRICES
VECTORS

WE'LL START OFF SLOW AND BUILD OUR WAY UP TO THE MORE ABSTRACT PARTS, OKAY?
DON'T WORRY, YOU'LL BE FINE.
SURE.

COMPLEX NUMBERS

Complex numbers are written in the form

$$a + b \cdot i$$

where a and b are real numbers and i is the *imaginary unit*, defined as $i = \sqrt{-1}$.

REAL NUMBERS			IMAGINARY NUMBERS
INTEGERS	RATIONAL NUMBERS* (NOT INTEGERS)	IRRATIONAL NUMBERS	
• Positive natural numbers • 0 • Negative natural numbers	• Terminating decimal numbers like 0.3 • Non-terminating decimal numbers like 0.333...	• Numbers like π and $\sqrt{2}$ whose decimals do not follow a pattern and repeat forever	• Complex numbers without a real component, like $0 + bi$, where b is a nonzero real number

* NUMBERS THAT CAN BE EXPRESSED IN THE FORM q / p (WHERE q AND p ARE INTEGERS, AND p IS NOT EQUAL TO ZERO) ARE KNOWN AS *RATIONAL NUMBERS*. INTEGERS ARE JUST SPECIAL CASES OF RATIONAL NUMBERS.

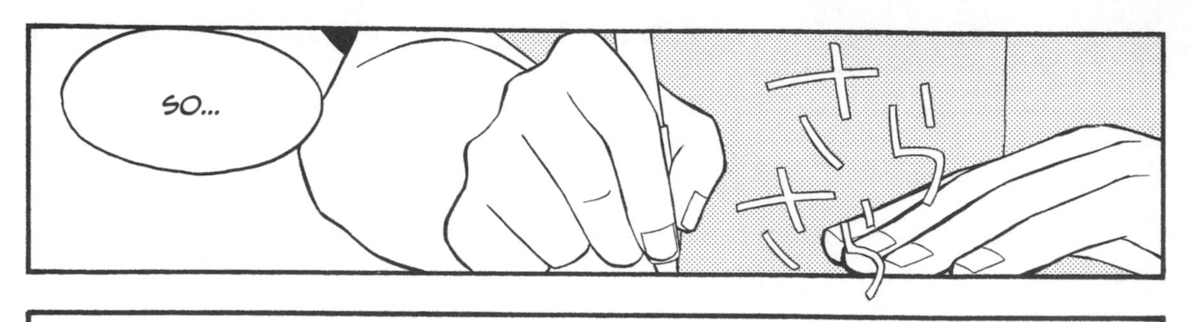
SO...
さら
さら

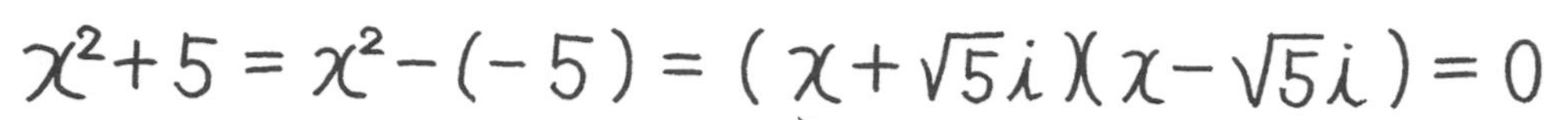
$x^2+5 = x^2-(-5) = (x+\sqrt{5}i)(x-\sqrt{5}i) = 0$

USING THIS NEW SYMBOL, THESE PREVIOUSLY UNSOLVABLE PROBLEMS SUDDENLY BECAME APPROACHABLE.

WHY WOULD YOU WANT TO SOLVE THEM IN THE FIRST PLACE? I DON'T REALLY SEE THE POINT.
I UNDERSTAND WHERE YOU'RE COMING FROM, BUT COMPLEX NUMBERS APPEAR PRETTY FREQUENTLY IN A VARIETY OF AREAS.

SIGH
ハァ
I'LL JUST HAVE TO GET USED TO THEM, I SUPPOSE...

DON'T WORRY! I THINK IT'D BE BETTER IF WE AVOIDED THEM FOR NOW SINCE THEY MIGHT MAKE IT HARDER TO UNDERSTAND THE REALLY IMPORTANT PARTS.
THANKS!

PROPOSITIONS

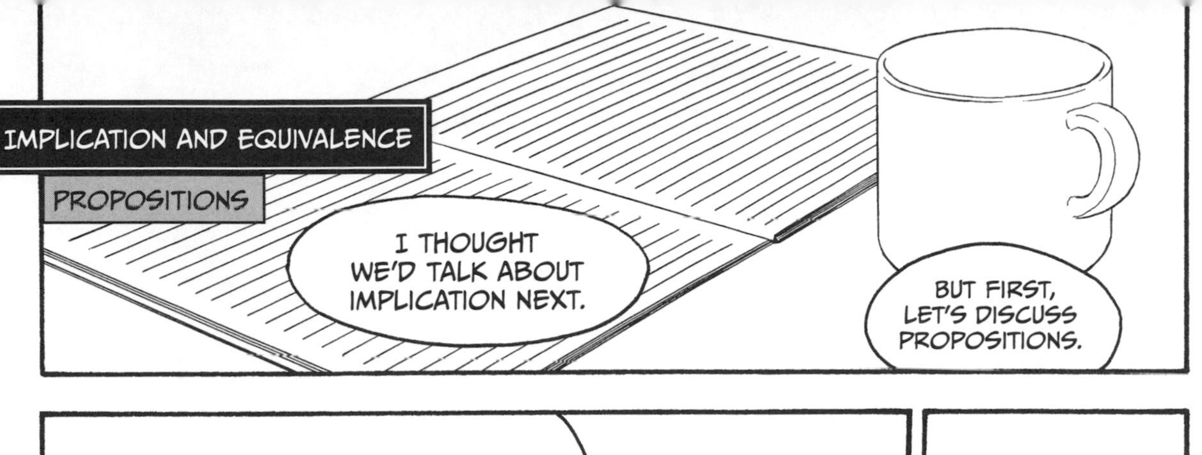

BUT IF WE LOOK AT ITS CONVERSE...

"IF THIS DISH CONTAINS PORK THEN IT IS A SCHNITZEL"

...IT IS NO LONGER NECESSARILY TRUE.

IN SITUATIONS WHERE WE KNOW THAT "IF P THEN Q" IS TRUE, BUT DON'T KNOW ANYTHING ABOUT ITS CONVERSE "IF Q THEN P"...

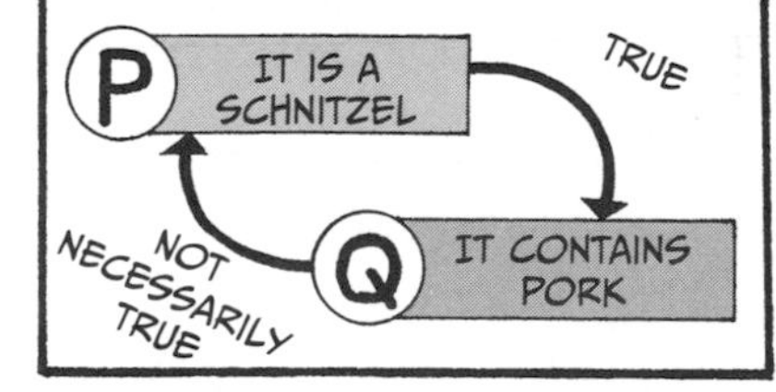

WE SAY THAT "P ENTAILS Q" AND THAT "Q COULD ENTAIL P."

IT IS A SCHNITZEL

ENTAILS

IT CONTAINS PORK

IT CONTAINS PORK

COULD ENTAIL

IT IS A SCHNITZEL

WHEN A PROPOSITION LIKE "IF P THEN Q" IS TRUE, IT IS COMMON TO WRITE IT WITH THE IMPLICATION SYMBOL, LIKE THIS:

$P \Rightarrow Q$

IF P THEN Q

$$P \Rightarrow Q$$

THIS IS A SCHNITZEL $\Rightarrow$ THIS DISH CONTAINS PORK

I THINK I GET IT.

EQUIVALENCE
IF BOTH "IF P THEN Q" AND "IF Q THEN P" ARE TRUE,
THAT IS, $P \Rightarrow Q$ AS WELL AS $Q \Rightarrow P$,

THEN P AND Q ARE *EQUIVALENT.*
SO IT'S LIKE THE IMPLICATION SYMBOLS POINT IN BOTH DIRECTIONS AT THE SAME TIME?
EXACTLY! IT'S KIND OF LIKE THIS.
DON'T WORRY. YOU'RE DUE FOR A GROWTH SPURT...
P
REIJI IS SHORTER THAN TETSUO.
Q
TETSUO IS TALLER THAN REIJI.

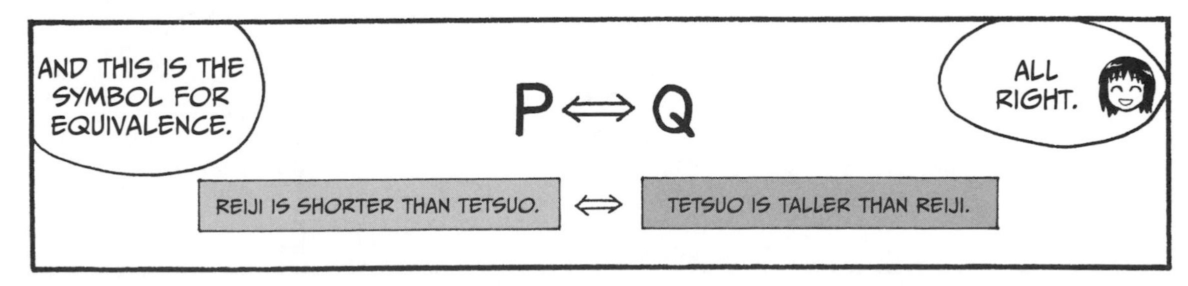
AND THIS IS THE SYMBOL FOR EQUIVALENCE.
$P \Leftrightarrow Q$
ALL RIGHT.
REIJI IS SHORTER THAN TETSUO. $\Leftrightarrow$ TETSUO IS TALLER THAN REIJI.

SET THEORY
SETS
ANOTHER IMPORTANT FIELD OF MATHEMATICS IS SET THEORY.
OH YEAH...I THINK WE COVERED THAT IN HIGH SCHOOL.
PROBABLY, BUT LET'S REVIEW IT ANYWAY.
SLIDE
JUST AS YOU MIGHT THINK, A *SET* IS A COLLECTION OF THINGS.

THE THINGS THAT MAKE UP THE SET ARE CALLED ITS *ELEMENTS* OR *OBJECTS*.
HEHE, OKAY.
THIS MIGHT GIVE YOU A GOOD IDEA OF WHAT I MEAN.

EXAMPLE 1

The set "Shikoku," which is the smallest of Japan's four islands, consists of these four elements:

- Kagawa-ken[1]
- Ehime-ken
- Kouchi-ken
- Tokushima-ken

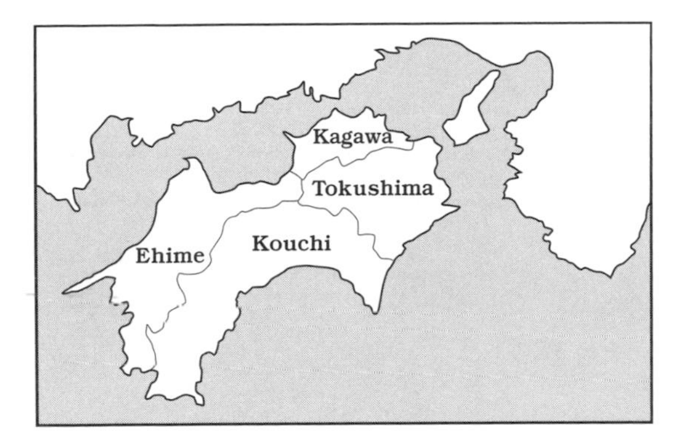

EXAMPLE 2

The set consisting of all even integers from 1 to 10 contains these five elements:

- 2
- 4
- 6
- 8
- 10

1. A Japanese *ken* is kind of like an American state.

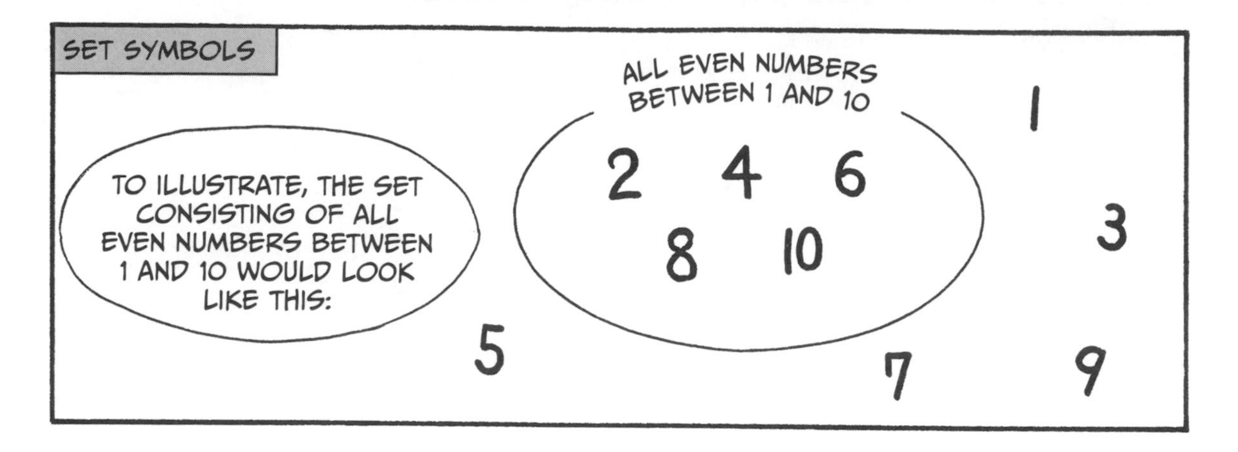

THESE ARE TWO COMMON WAYS TO WRITE OUT THAT SET:

$\{2, 4, 6, 8, 10\}$ $\{2n \mid n = 1, 2, 3, 4, 5\}$

MMM...

WITH THAT IN MIND, OUR DEFINITION NOW LOOKS LIKE THIS:

$X = \{2, 4, 6, 8, 10\}$

$X = \{2n \mid n = 1, 2, 3, 4, 5\}$

THIS IS A GOOD WAY TO EXPRESS THAT "THE ELEMENT x BELONGS TO THE SET X."

$x \in X$

LET'S SAY THAT ALL ELEMENTS OF A SET X ALSO BELONG TO A SET Y.

X IS A *SUBSET* OF Y IN THIS CASE.

SET Y (JAPAN)

HOKKAIDOU
AOMORI-KEN
IWATE-KEN
MIYAGI-KEN
AKITA-KEN
YAMAGATA-KEN
FUKUSHIMA-KEN
IBARAKI-KEN
TOCHIGI-KEN
GUNMA-KEN
SAITAMA-KEN
CHIBA-KEN
TOUKYOU-TO
KANAGAWA-KEN
NIIGATA-KEN
TOYAMA-KEN
ISHIKAWA-KEN
FUKUI-KEN
YAMANASHI-KEN
NAGANO-KEN
GIFU-KEN
SHIZUOKA-KEN
AICHI-KEN
MIE-KEN
SHIGA-KEN
KYOUTO-FU
OOSAKA-FU
HYOUGO-KEN
NARA-KEN
WAKAYAMA-KEN
TOTTORI-KEN
SHIMANE-KEN
OKAYAMA-KEN
HIROSHIMA-KEN
YAMAGUCHI-KEN
FUKUOKA-KEN
SAGA-KEN
NAGASAKI-KEN
KUMAMOTO-KEN
OOITA-KEN
MIYAZAKI-KEN
KAGOSHIMA-KEN
OKINAWA-KEN

SET X (SHIKOKU)

KAGAWA-KEN
EHIME-KEN
KOUCHI-KEN
TOKUSHIMA-KEN

EXAMPLE 1

Suppose we have two sets X and Y:

$X = \{ 4, 10 \}$
$Y = \{ 2, 4, 6, 8, 10 \}$

X is a subset of Y, since all elements in X also exist in Y.

EXAMPLE 2

Suppose we switch the sets:

$X = \{ 2, 4, 6, 8, 10 \}$
$Y = \{ 4, 10 \}$

Since all elements in X don't exist in Y, X is no longer a subset of Y.

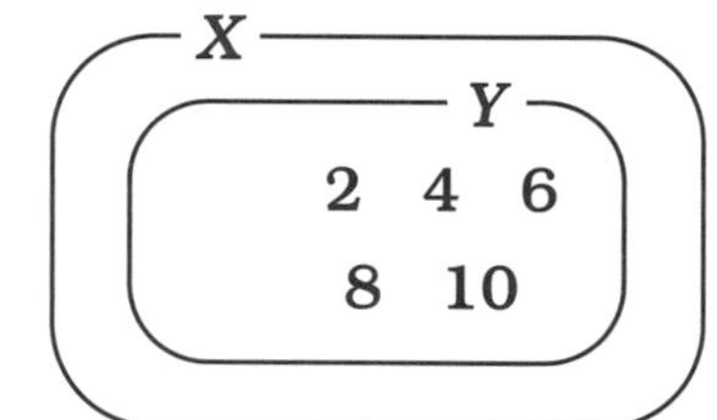

EXAMPLE 3

Suppose we have two equal sets instead:

$X = \{ 2, 4, 6, 8, 10 \}$
$Y = \{ 2, 4, 6, 8, 10 \}$

In this case, both sets are subsets of each other. So X is a subset of Y, and Y is a subset of X.

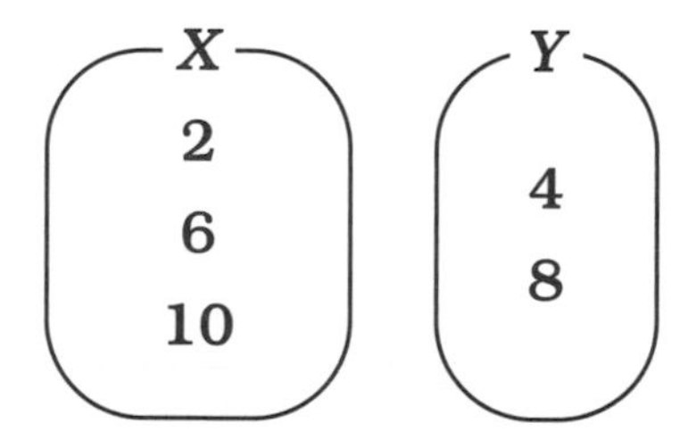

EXAMPLE 4

Suppose we have the two following sets:

$X = \{ 2, 6, 10 \}$
$Y = \{ 4, 8 \}$

In this case neither X nor Y is a subset of the other.

FUNCTIONS

I THOUGHT WE'D TALK ABOUT FUNCTIONS AND THEIR RELATED CONCEPTS NEXT.
FUNCTIONS

IT'S ALL PRETTY ABSTRACT, BUT YOU'LL BE FINE AS LONG AS YOU TAKE YOUR TIME AND THINK HARD ABOUT EACH NEW IDEA.
GOT IT.

LET'S START BY DEFINING THE CONCEPT ITSELF.
SOUNDS GOOD.

IMAGINE THE FOLLOWING SCENARIO:
CAPTAIN ICHINOSE, IN A PLEASANT MOOD, DECIDES TO TREAT US FRESHMEN TO LUNCH.
SO WE FOLLOW HIM TO RESTAURANT A.
FOLLOW ME!

THIS IS THE RESTAURANT MENU.

UDON 500 YEN

CURRY 700 YEN

BREADED PORK 1000 YEN

BROILED EEL 1500 YEN

BUT THERE IS A CATCH, OF COURSE.
WHAT DO YOU MEAN?

SINCE HE'S THE ONE PAYING, HE GETS A SAY IN ANY AND ALL ORDERS.
?

KIND OF LIKE THIS:

WE WOULDN'T REALLY BE ABLE TO SAY NO IF HE TOLD US TO ORDER THE CHEAPEST DISH, RIGHT?
UDON FOR EVERYONE!
YURINO
YOSHIDA
YAJIMA
TOMIYAMA
UDON
CURRY
BREADED PORK
BROILED EEL
OR SAY, IF HE JUST TOLD US ALL TO ORDER DIFFERENT THINGS.
ORDER DIFFERENT STUFF!
YURINO
YOSHIDA
YAJIMA
TOMIYAMA
UDON
CURRY
BREADED PORK
BROILED EEL

EVEN IF HE TOLD US TO ORDER OUR FAVORITES, WE WOULDN'T REALLY HAVE A CHOICE. THIS MIGHT MAKE US THE MOST HAPPY, BUT THAT DOESN'T CHANGE THE FACT THAT WE HAVE TO OBEY HIM.

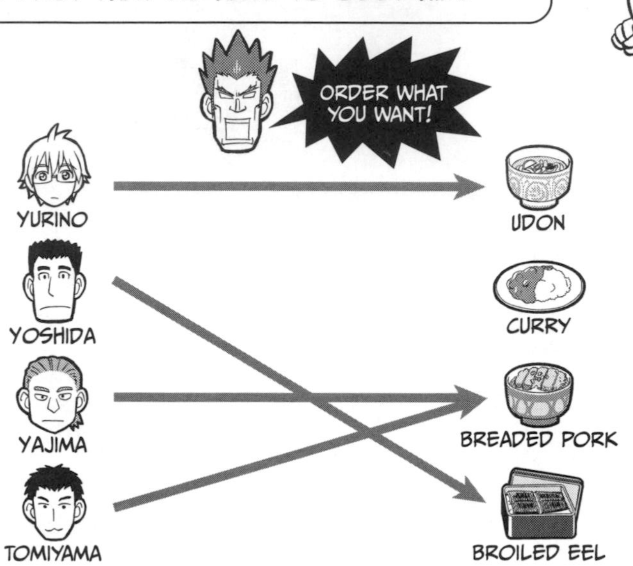

YOU COULD SAY THAT THE CAPTAIN'S ORDERING GUIDELINES ARE LIKE A "RULE" THAT BINDS ELEMENTS OF X TO ELEMENTS OF Y.

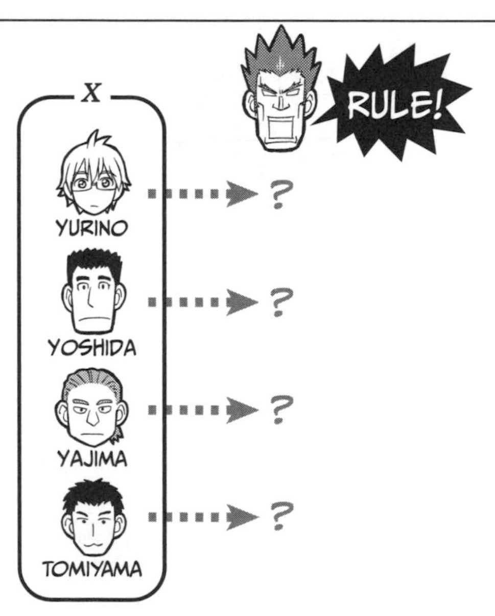

THIS IS HOW WE WRITE IT:

$$X \xrightarrow{f} Y \quad \text{or} \quad f : X \longrightarrow Y$$

CLUB MEMBER $\xrightarrow{\text{RULE}}$ MENU OR RULE : CLUB MEMBER $\longrightarrow$ MENU

FUNCTIONS

A rule that binds elements of the set X to elements of the set Y is called "a *function* from X to Y." X is usually called the *domain* and Y the *co-domain* or *target set* of the function.

IMAGES
NEXT UP ARE IMAGES.
IMAGES?

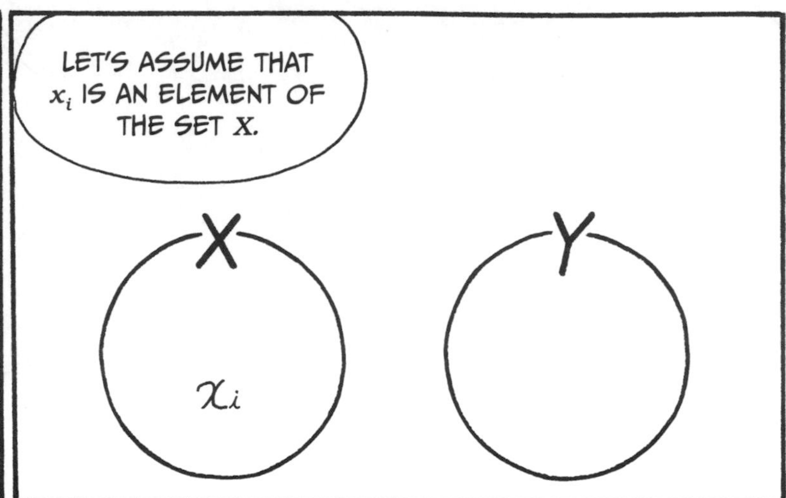
LET'S ASSUME THAT x_i IS AN ELEMENT OF THE SET X.
X
Y
x_i

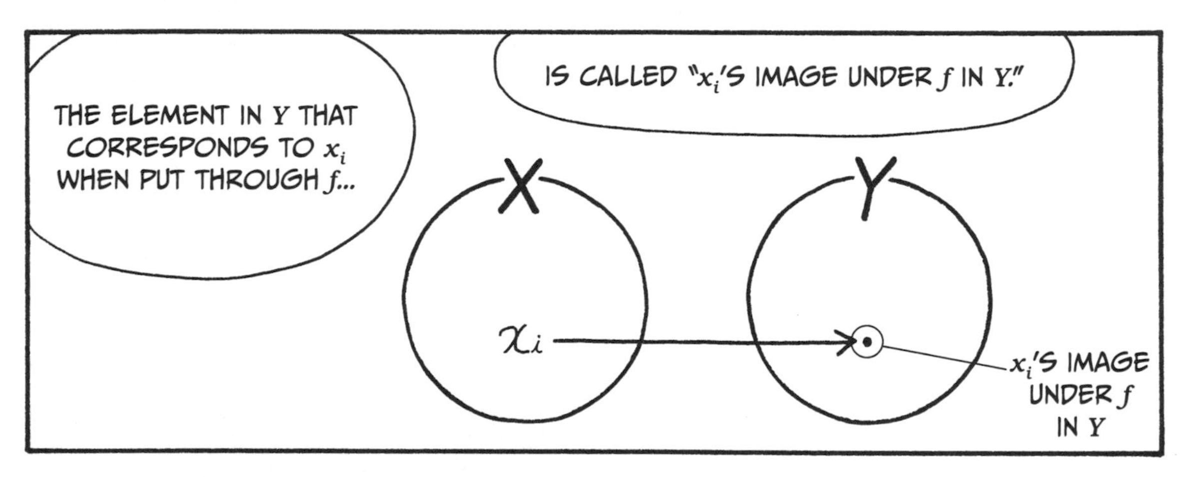
THE ELEMENT IN Y THAT CORRESPONDS TO x_i WHEN PUT THROUGH f...
IS CALLED "x_i'S IMAGE UNDER f IN Y."
X
Y
x_i
x_i'S IMAGE UNDER f IN Y

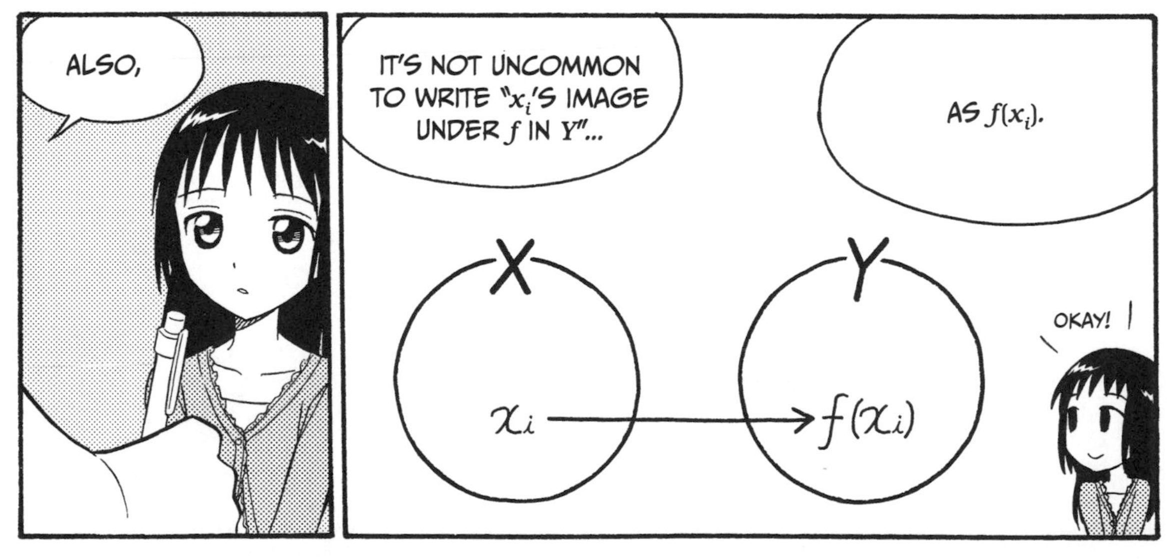
ALSO,
IT'S NOT UNCOMMON TO WRITE "x_i'S IMAGE UNDER f IN Y"...
AS $f(x_i)$.
X
Y
x_i
$f(x_i)$
OKAY!

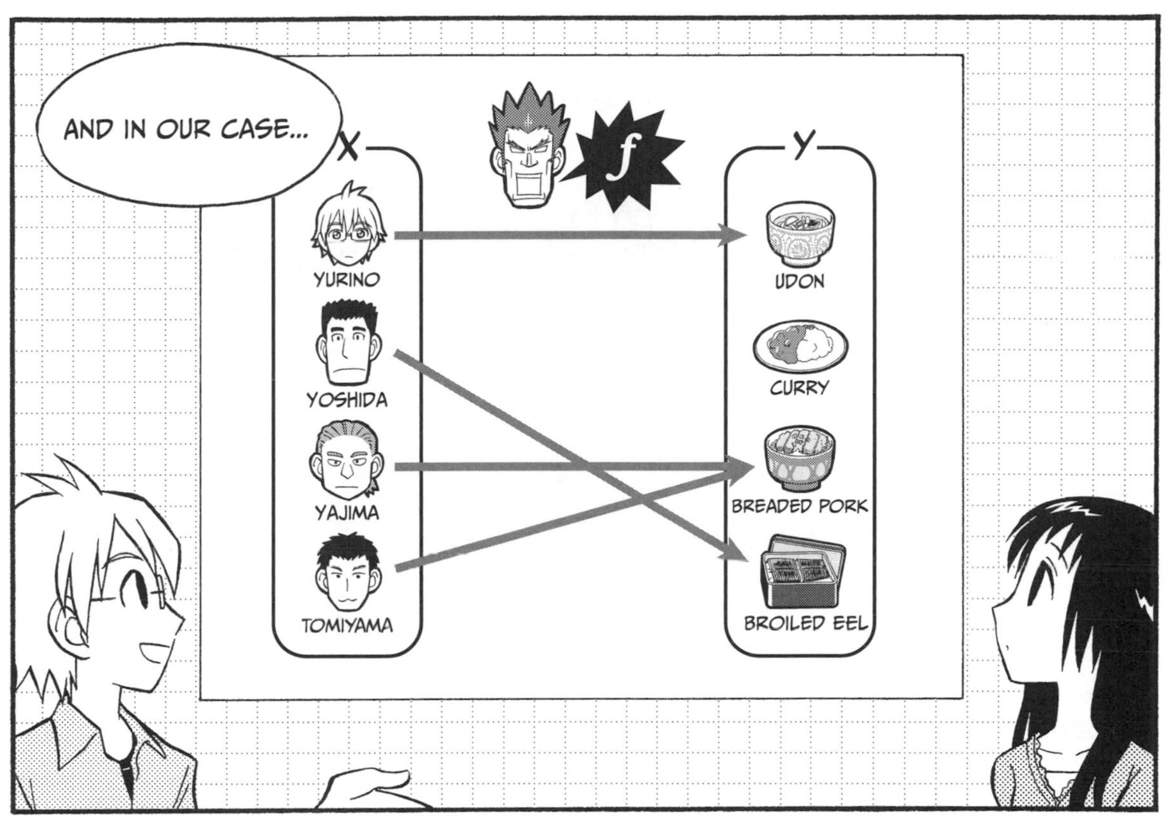

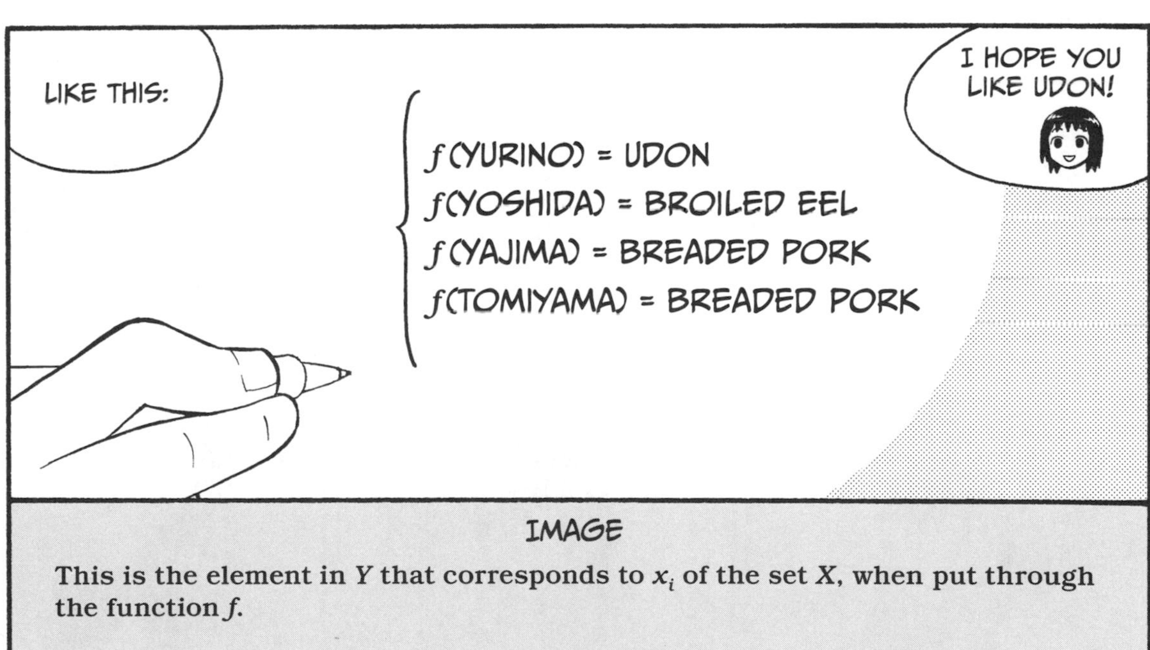

IMAGE

This is the element in Y that corresponds to x_i of the set X, when put through the function f.

BY THE WAY, DO YOU REMEMBER THIS TYPE OF FORMULA FROM YOUR HIGH SCHOOL YEARS?
?

OH... YEAH, SURE.
$f(x)=2x-1$

DIDN'T YOU EVER WONDER WHY...

...THEY ALWAYS USED THIS WEIRD SYMBOL $f(x)$ WHERE THEY COULD HAVE USED SOMETHING MUCH SIMPLER LIKE y INSTEAD?

"LIKE WHATEVER! ANYWAYS, SO IF I WANT TO SUBSTITUTE WITH 2 IN THIS FORMULA, I'M SUPPOSED TO WRITE $f(2)$ AND..."
INSIDE MISA'S BRAIN

ACTUALLY... I HAVE!

WELL, HERE'S WHY.
WHAT $f(x) = 2x - 1$ REALLY MEANS IS THIS:
THE FUNCTION f IS A RULE THAT SAYS:
"THE ELEMENT x OF THE SET X GOES TOGETHER WITH THE ELEMENT $2x - 1$ IN THE SET Y."

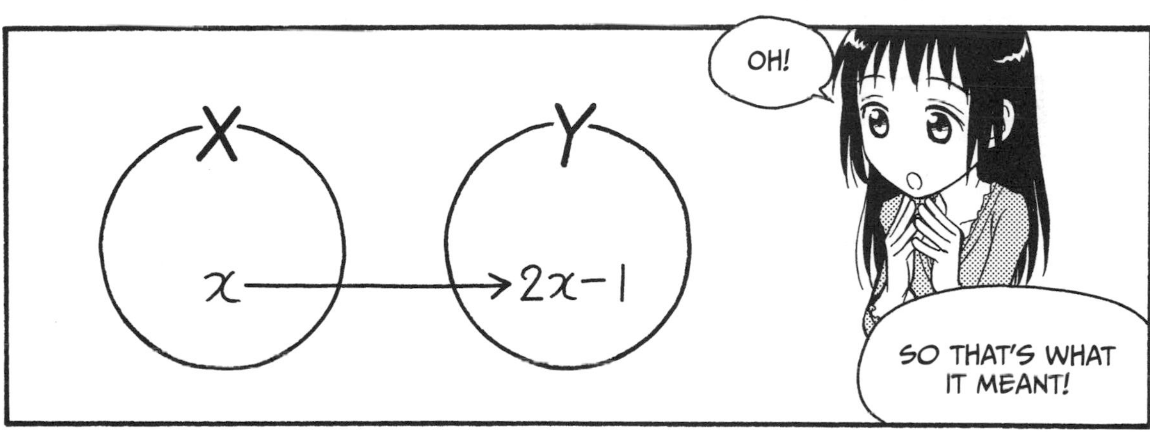
OH!
X
Y
x
2x−1
SO THAT'S WHAT IT MEANT!

SIMILARLY, $f(2)$ IMPLIES THIS:
I THINK I'M STARTING TO GET IT.
The image of 2 under the function f is $2 \cdot 2 - 1$.
X
Y
2
2 · 2-1
SO WE WERE USING FUNCTIONS IN HIGH SCHOOL TOO?
EXACTLY.

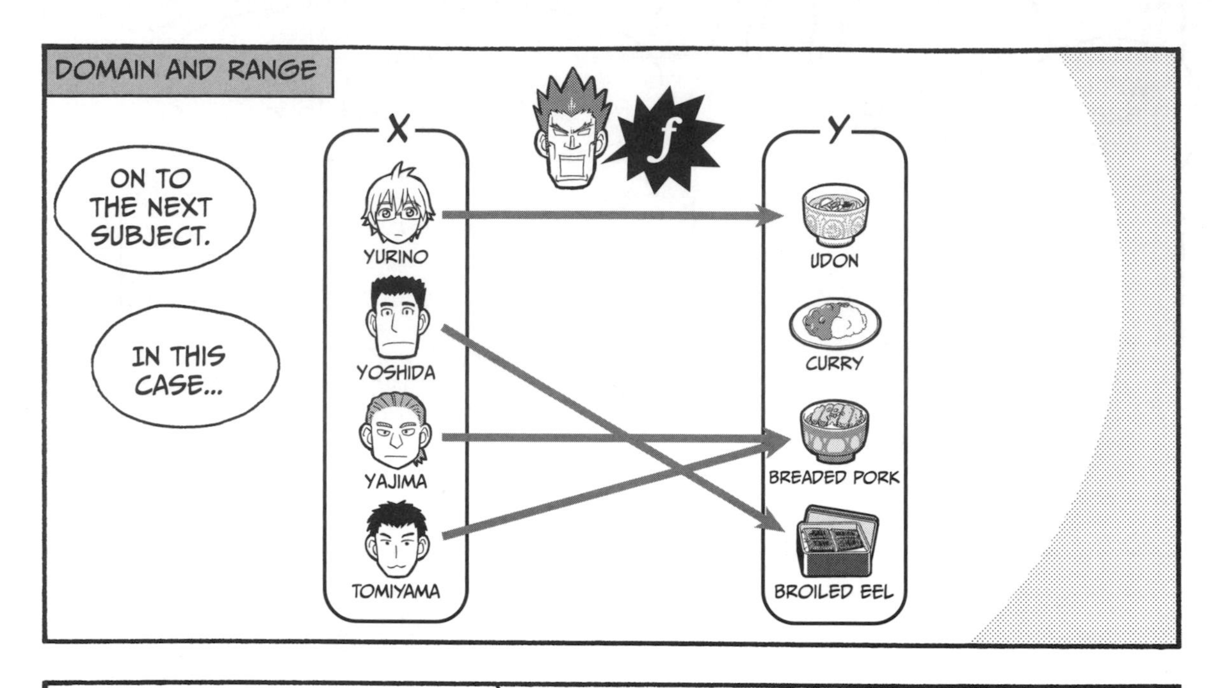

* THE TERM *IMAGE* IS USED HERE TO DESCRIBE THE *SET* OF ELEMENTS IN Y THAT ARE THE IMAGE OF AT LEAST ONE ELEMENT IN X.

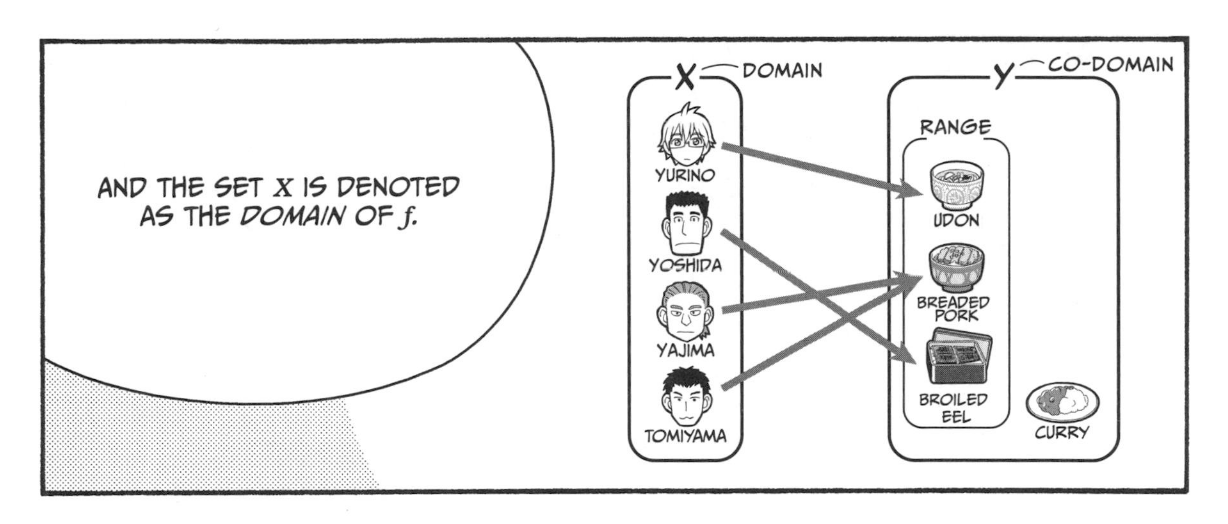

RANGE AND CO-DOMAIN

The set that encompasses the function f's image $\{f(x_1), f(x_2), \ldots, f(x_n)\}$ is called the *range* of f, and the (possibly larger) set being mapped into is called its *co-domain*.

The relationship between the range and the co-domain Y is as follows:

$$\{f(x_1), f(x_2), \ldots, f(x_n)\} \subset Y$$

In other words, a function's range is a subset of its co-domain. In the special case where all elements in Y are an image of some element in X, we have

$$\{f(x_1), f(x_2), \ldots, f(x_n)\} = Y$$

LET'S SAY OUR KARATE CLUB DECIDES TO HAVE A PRACTICE MATCH WITH ANOTHER CLUB...

AND THAT THE CAPTAIN'S MAPPING FUNCTION f IS "FIGHT THAT GUY."

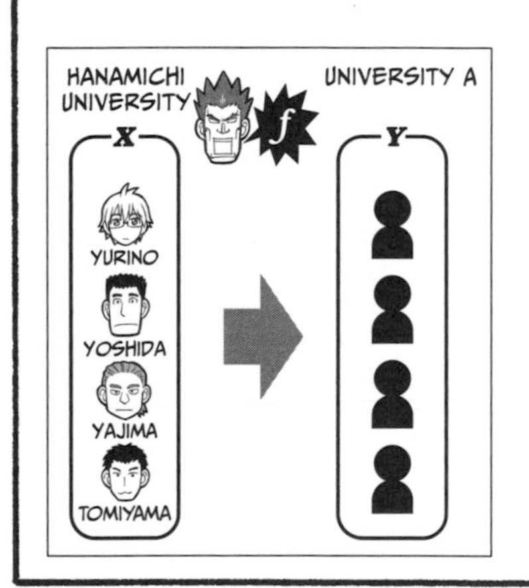

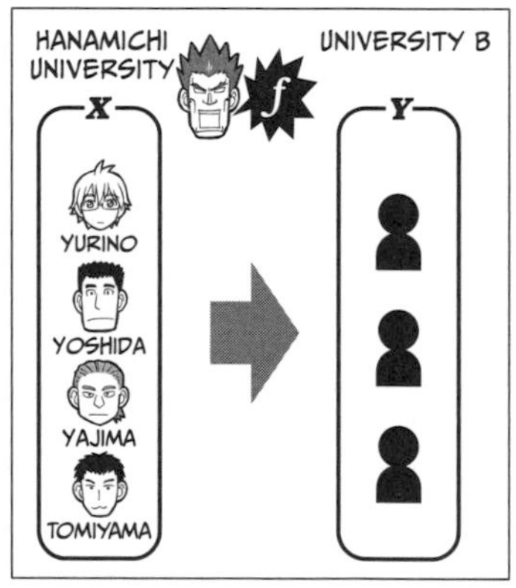

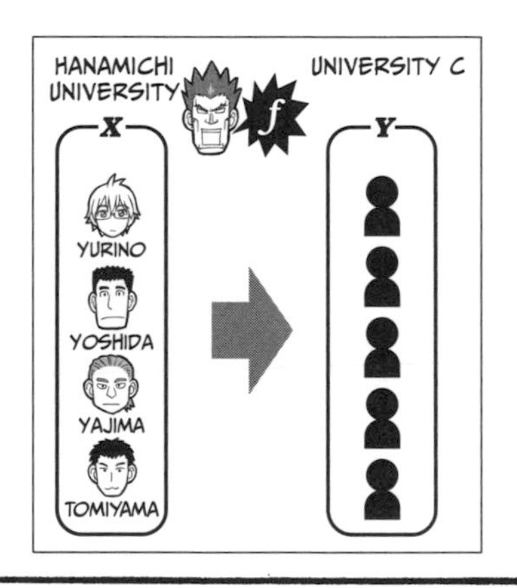

ONTO FUNCTIONS

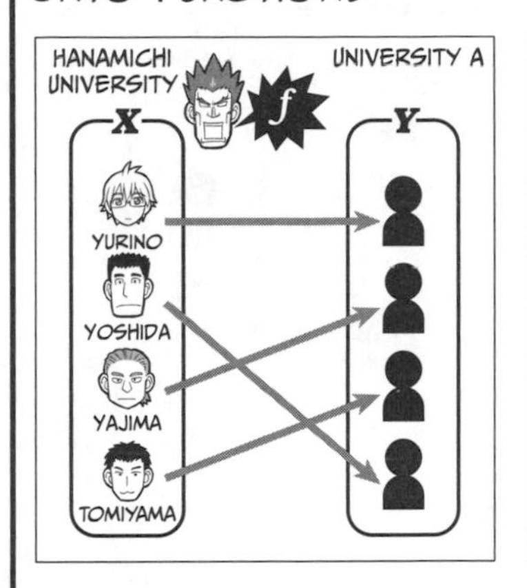

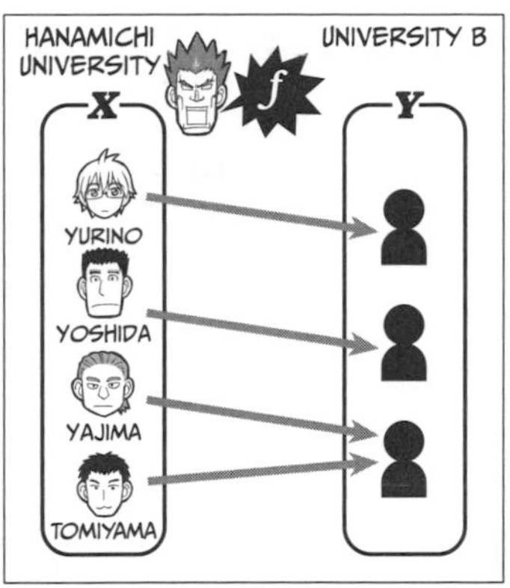

A FUNCTION IS *ONTO* IF ITS IMAGE IS EQUAL TO ITS CO-DOMAIN. THIS MEANS THAT ALL THE ELEMENTS IN THE CO-DOMAIN OF AN ONTO FUNCTION ARE BEING MAPPED ONTO.

ONE-TO-ONE FUNCTIONS

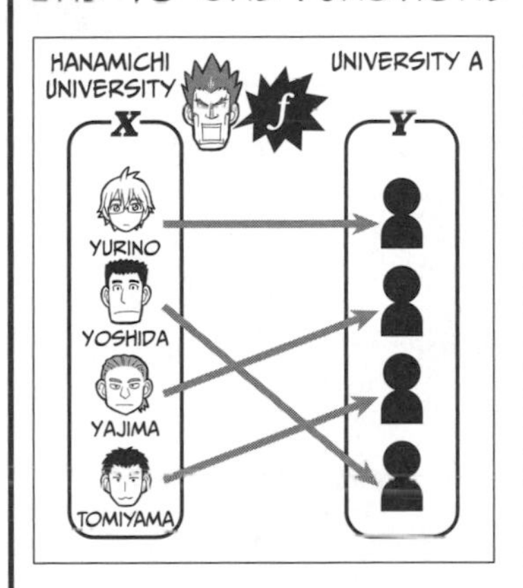

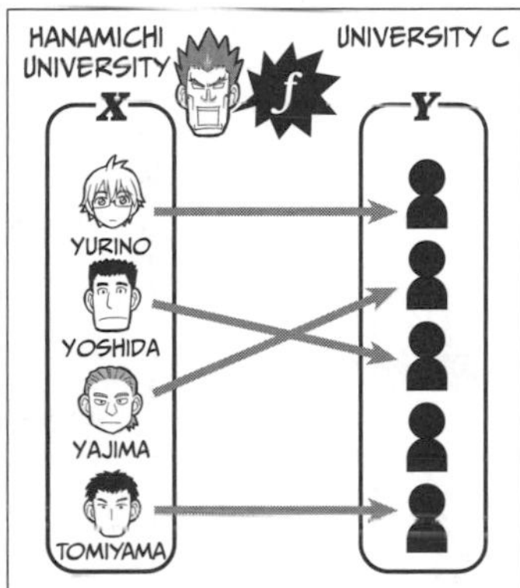

IF $x_i \neq x_j$ LEADS TO $f(x_i) \neq f(x_j)$, WE SAY THAT THE FUNCTION IS *ONE-TO-ONE*. THIS MEANS THAT NO ELEMENT IN THE CO-DOMAIN CAN BE MAPPED ONTO MORE THAN ONCE.

ONE-TO-ONE AND ONTO FUNCTIONS

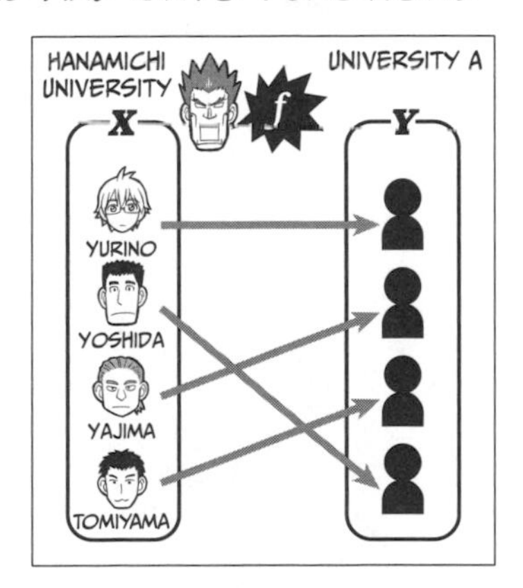

IT'S ALSO POSSIBLE FOR A FUNCTION TO BE BOTH ONTO AND ONE-TO-ONE. SUCH A FUNCTION CREATES A "BUDDY SYSTEM" BETWEEN THE ELEMENTS OF THE DOMAIN AND CO-DOMAIN. EACH ELEMENT HAS ONE AND ONLY ONE "PARTNER."

INVERSE FUNCTIONS
NOW WE HAVE INVERSE FUNCTIONS.
INVERSE?
THIS TIME WE'RE GOING TO LOOK AT THE OTHER TEAM CAPTAIN'S ORDERS AS WELL.
X
Y
g
YURINO
YOSHIDA
YAJIMA
TOMIYAMA

HANAMICHI UNIVERSITY
UNIVERSITY A
f
X
Y
YURINO
YOSHIDA
YAJIMA
TOMIYAMA
HANAMICHI UNIVERSITY
UNIVERSITY A
g
X
Y
YURINO
YOSHIDA
YAJIMA
TOMIYAMA
WE SAY THAT THE FUNCTION g IS f'S *INVERSE* WHEN THE TWO CAPTAINS' ORDERS COINCIDE LIKE THIS.
I SEE.

TO SPECIFY EVEN FURTHER...

f IS AN INVERSE OF g IF THESE TWO RELATIONS HOLD.

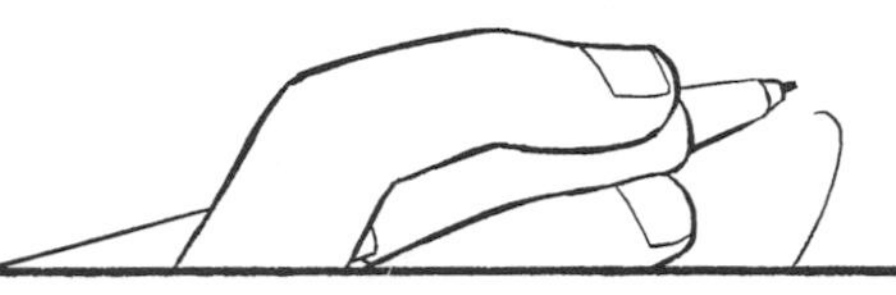

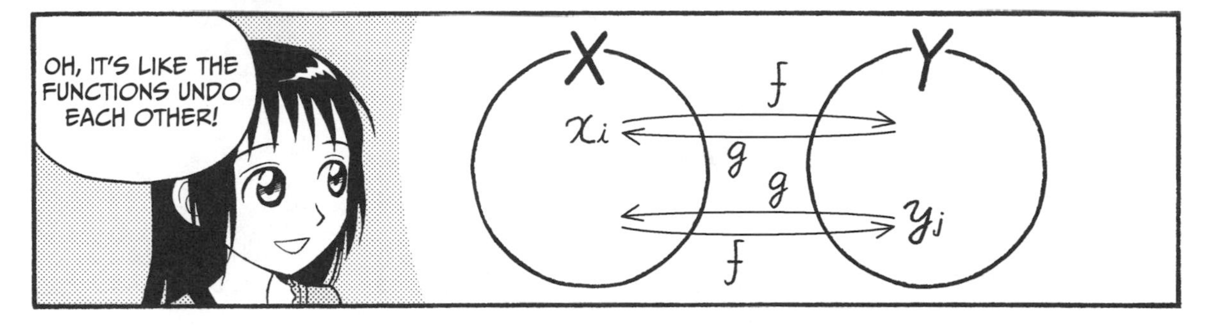

THIS IS THE SYMBOL USED TO INDICATE INVERSE FUNCTIONS.

$$X \xrightarrow{f^{-1}} Y$$

OR

$$f^{-1}: X \rightarrow Y$$

YOU RAISE IT TO −1, RIGHT?

THERE IS ALSO A CONNECTION BETWEEN ONE-TO-ONE AND ONTO FUNCTIONS AND INVERSE FUNCTIONS. HAVE A LOOK AT THIS.

THE FUNCTION f HAS AN INVERSE. $\Longleftrightarrow$ THE FUNCTION f IS ONE-TO-ONE AND ONTO.

SO IF IT'S ONE-TO-ONE AND ONTO, IT HAS AN INVERSE, AND VICE VERSA. GOT IT!

LINEAR TRANSFORMATIONS

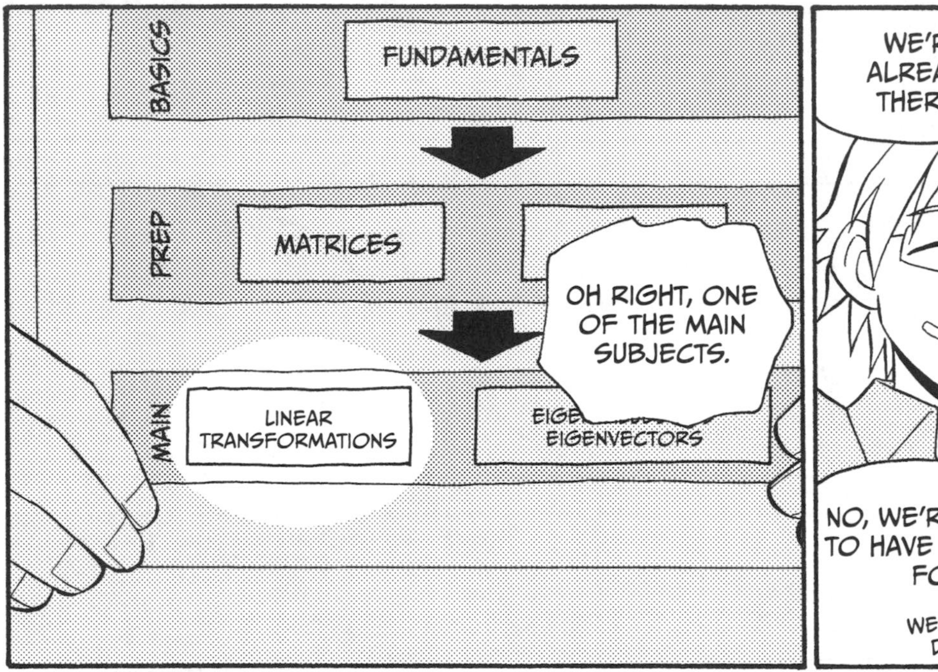

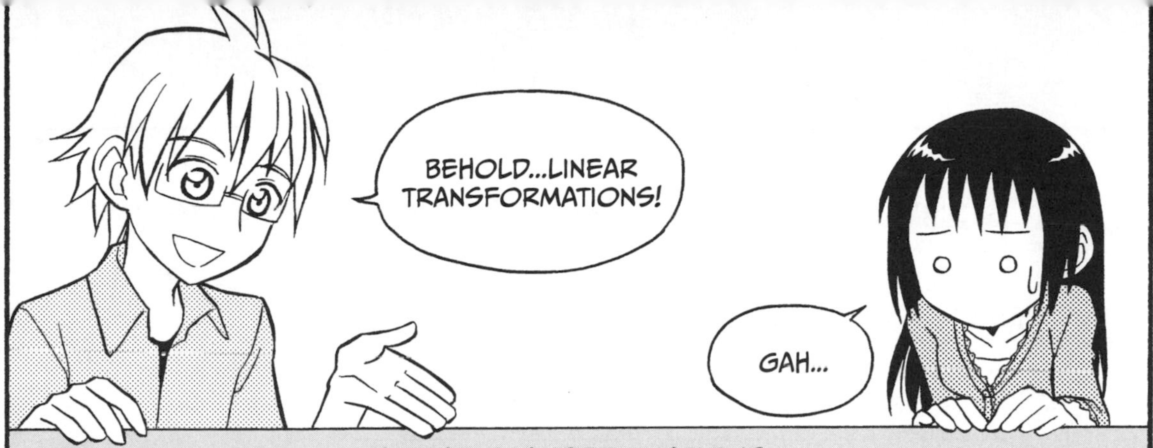

LINEAR TRANSFORMATIONS

Let x_i and x_j be two arbitrary elements of the set X, c be any real number, and f be a function from X to Y. f is called a *linear transformation* from X to Y if it satisfies the following two conditions:

❶ $f(x_i) + f(x_j) = f(x_i + x_j)$

❷ $cf(x_i) = f(cx_i)$

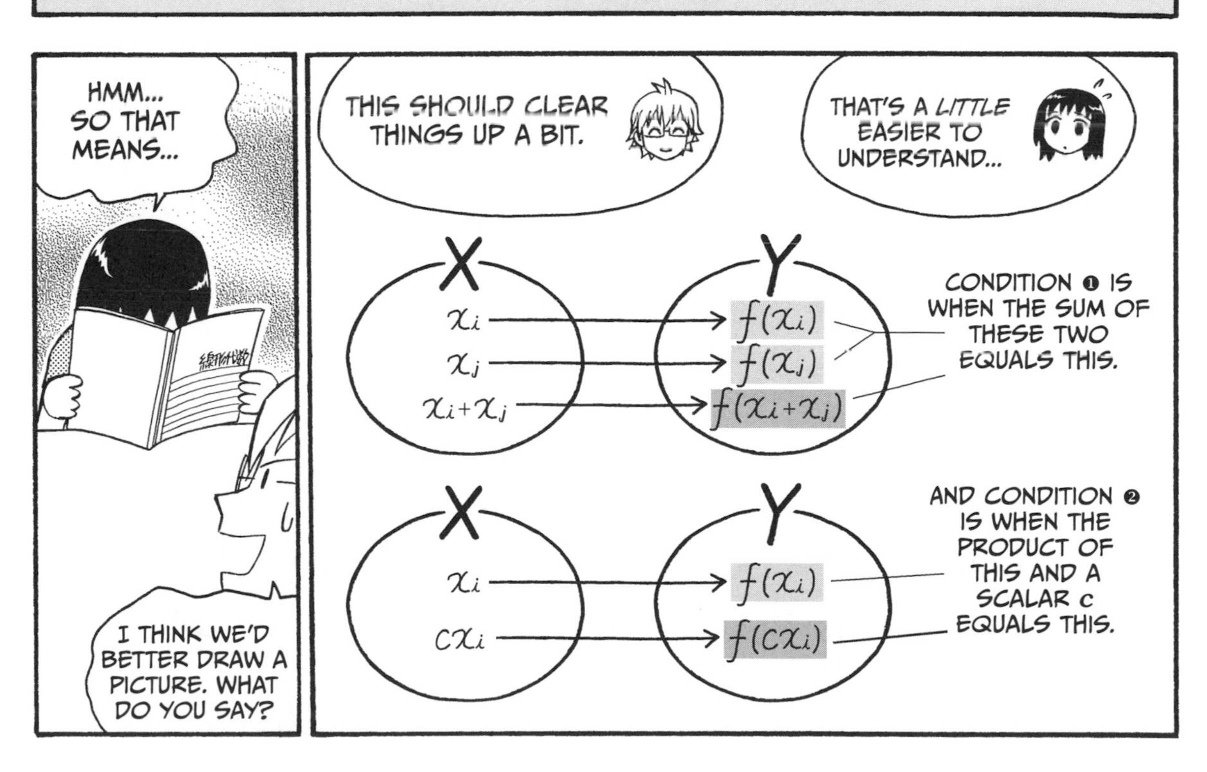

AN EXAMPLE OF A LINEAR TRANSFORMATION

The function $f(x) = 2x$ is a linear transformation. This is because it satisfies both ❶ and ❷, as you can see in the table below.

Condition ❶	$\begin{cases} f(x_i) + f(x_j) = 2x_i + 2x_j \\ f(x_i + x_j) = 2(x_i + x_j) = 2x_i + 2x_j \end{cases}$
Condition ❷	$\begin{cases} cf(x_i) = c(2x_i) = 2cx_i \\ f(cx_i) = 2(cx_i) = 2cx_i \end{cases}$

AN EXAMPLE OF A FUNCTION THAT IS NOT A LINEAR TRANSFORMATION

The function $f(x) = 2x - 1$ is not a linear transformation. This is because it satisfies neither ❶ nor ❷, as you can see in the table below.

Condition ❶	$\begin{cases} f(x_i) + f(x_j) = (2x_i - 1) + (2x_j - 1) = 2x_i + 2x_j - 2 \\ f(x_i + x_j) = 2(x_i + x_j) - 1 = 2x_i + 2x_j - 1 \end{cases}$
Condition ❷	$\begin{cases} cf(x_i) = c(2x_i - 1) = 2cx_i - c \\ f(cx_i) = 2(cx_i) - 1 = 2cx_i - 1 \end{cases}$

REIJI?
UM...
DO YOU ALWAYS EAT LUNCH IN THE SCHOOL CAFETERIA?
I LIVE ALONE, AND I'M NOT THAT GOOD AT COOKING, SO MOST OF THE TIME, YEAH...
WELL, NEXT TIME WE MEET YOU WON'T HAVE TO. I'M MAKING YOU LUNCH!
え!?
DON'T WORRY ABOUT IT! TETSUO IS STILL HELPING ME OUT AND ALL!
YOU DON'T WANT ME TO?

NO, THAT'S NOT IT, IT'S JUST...
UH...

ON SECOND THOUGHT, I'D LOVE FOR YOU TO...

...MAKE ME LUNCH.

GREAT!

I MAKE A LOT OF THEM FOR MY BROTHER TOO, YOU KNOW—STAMINA LUNCHES.
OH...HOW LOVELY.

COMBINATIONS AND PERMUTATIONS

I thought the best way to explain combinations and permutations would be to give a concrete example.

I'll start by explaining the PROBLEM, then I'll establish a good WAY OF THINKING, and finally I'll present a SOLUTION.

PROBLEM

Reiji bought a CD with seven different songs on it a few days ago. Let's call the songs A, B, C, D, E, F, and G. The following day, while packing for a car trip he had planned with his friend Nemoto, it struck him that it might be nice to take the songs along to play during the drive. But he couldn't take all of the songs, since his taste in music wasn't very compatible with Nemoto's. After some deliberation, he decided to make a new CD with only three songs on it from the original seven.

Questions:

1. In how many ways can Reiji select three songs from the original seven?
2. In how many ways can the three songs be arranged?
3. In how many ways can a CD be made, where three songs are chosen from a pool of seven?

WAY OF THINKING

It is possible to solve question 3 by dividing it into these two subproblems:

1. Choose three songs out of the seven possible ones.
2. Choose an order in which to play them.

As you may have realized, these are the first two questions. The solution to question 3, then, is as follows:

SOLUTION TO QUESTION 1 ·	SOLUTION TO QUESTION 2 =	SOLUTION TO QUESTION 3
In how many ways can Reiji select three songs from the original seven?	In how many ways can the three songs be arranged?	In how many ways can a CD be made, where three songs are chosen from a pool of seven?

SOLUTION

1. **In how many ways can Reiji select three songs from the original seven?**

All 35 different ways to select the songs are in the table below. Feel free to look them over.

Pattern	Songs
Pattern 1	A and B and C
Pattern 2	A and B and D
Pattern 3	A and B and E
Pattern 4	A and B and F
Pattern 5	A and B and G
Pattern 6	A and C and D
Pattern 7	A and C and E
Pattern 8	A and C and F
Pattern 9	A and C and G
Pattern 10	A and D and E
Pattern 11	A and D and F
Pattern 12	A and D and G
Pattern 13	A and E and F
Pattern 14	A and E and G
Pattern 15	A and F and G
Pattern 16	B and C and D
Pattern 17	B and C and E
Pattern 18	B and C and F
Pattern 19	B and C and G
Pattern 20	B and D and E
Pattern 21	B and D and F
Pattern 22	B and D and G
Pattern 23	B and E and F
Pattern 24	B and E and G
Pattern 25	B and F and G
Pattern 26	C and D and E
Pattern 27	C and D and F
Pattern 28	C and D and G
Pattern 29	C and E and F
Pattern 30	C and E and G
Pattern 31	C and F and G
Pattern 32	D and E and G
Pattern 33	D and E and G
Pattern 34	D and F and G
Pattern 35	E and F and G

Choosing k among n items without considering the order in which they are chosen is called a *combination*. The number of different ways this can be done is written by using the binomial coefficient notation:

$$\binom{n}{k}$$

which is read "n choose k."

In our case,

$$\binom{7}{3} = 35$$

2. In how many ways can the three songs be arranged?

Let's assume we chose the songs A, B, and C. This table illustrates the 6 different ways in which they can be arranged:

Song 1	Song 2	Song 3
A	B	C
A	C	B
B	A	C
B	C	A
C	A	B
C	B	A

Suppose we choose B, E, and G instead:

Song 1	Song 2	Song 3
B	E	G
B	G	E
E	B	G
E	G	B
G	B	E
G	E	B

Trying a few other selections will reveal a pattern: The number of possible arrangements does not depend on which three elements we choose—there are always six of them. Here's why:

Our result (6) can be rewritten as $3 \cdot 2 \cdot 1$, which we get like this:

1. We start out with all three songs and can choose any one of them as our first song.
2. When we're picking our second song, only two remain to choose from.
3. For our last song, we're left with only one choice.

This gives us 3 possibilities · 2 possibilities · 1 possibility = 6 possibilities.

3. In how many ways can a CD be made, where three songs are chosen from a pool of seven?

The different possible patterns are

The number of ways to choose three songs from seven	$\cdot$	The number of ways the three songs can be arranged

$$= \binom{7}{3} \cdot 6$$

$$= 35 \cdot 6$$

$$= 210$$

This means that there are 210 different ways to make the CD.

Choosing three from seven items in a certain order creates a *permutation* of the chosen items. The number of possible permutations of k objects chosen among n objects is written as

$${}_nP_k$$

In our case, this comes to

$${}_7P_3 = 210$$

The number of ways n objects can be chosen among n possible ones is equal to n-factorial:

$${}_nP_n = n! = n \cdot (n-1) \cdot (n-2) \cdot \ldots \cdot 2 \cdot 1$$

For instance, we could use this if we wanted to know how many different ways seven objects can be arranged. The answer is

$$7! = 7 \cdot 6 \cdot 5 \cdot 4 \cdot 3 \cdot 2 \cdot 1 = 5040$$

I've listed all possible ways to choose three songs from the seven original ones (A, B, C, D, E, F, and G) in the table below.

	Song 1	Song 2	Song 3
Pattern 1	A	B	C
Pattern 2	A	B	D
Pattern 3	A	B	E
...	...	...	...
Pattern 30	A	G	F
Pattern 31	B	A	C
...	...	...	...
Pattern 60	B	G	F
Pattern 61	C	A	B
...	...	...	...
Pattern 90	C	G	F
Pattern 91	D	A	B
...	...	...	...
Pattern 120	D	G	F
Pattern 121	E	A	B
...	...	...	...
Pattern 150	E	G	F
Pattern 151	F	A	B
...	...	...	...
Pattern 180	F	G	E
Pattern 181	G	A	B
...	...	...	...
Pattern 209	G	E	F
Pattern 210	G	F	E

We can, analogous to the previous example, rewrite our problem of counting the different ways in which to make a CD as $7 \cdot 6 \cdot 5 = 210$. Here's how we get those numbers:

1. We can choose any of the **7** songs A, B, C, D, E, F, and G as our first song.
2. We can then choose any of the **6** remaining songs as our second song.
3. And finally we choose any of the now **5** remaining songs as our last song.

The definition of the binomial coefficient is as follows:

$$\binom{n}{r} = \frac{n \cdot (n-1) \cdots (n-(r-1))}{r \cdot (r-1) \cdots 1} = \frac{n \cdot (n-1) \cdots (n-r+1)}{r \cdot (r-1) \cdots 1}$$

Notice that

$$\begin{aligned}
\binom{n}{r} &= \frac{n \cdot (n-1) \cdots (n-(r-1))}{r \cdot (r-1) \cdots 1} \\
&= \frac{n \cdot (n-1) \cdots (n-(r-1))}{r \cdot (r-1) \cdots 1} \cdot \frac{(n-r) \cdot (n-r+1) \cdots 1}{(n-r) \cdot (n-r+1) \cdots 1} \\
&= \frac{n \cdot (n-1) \cdots (n-(r-1)) \cdot (n-r) \cdot (n-r+1) \cdots 1}{(r \cdot (r-1) \cdots 1) \cdot ((n-r) \cdot (n-r+1) \cdots 1)} \\
&= \frac{n!}{r! \cdot (n-r)!}
\end{aligned}$$

Many people find it easier to remember the second version:

$$\binom{n}{r} = \frac{n!}{r! \cdot (n-r)!}$$

We can rewrite question 3 (how many ways can the CD be made?) like this:

$${}_7P_3 = \binom{7}{3} \cdot 6 = \binom{7}{3} \cdot 3! = \frac{7!}{3! \cdot 4!} \cdot 3! = \frac{7!}{4!} = \frac{7 \cdot 6 \cdot 5 \cdot 4 \cdot 3 \cdot 2 \cdot 1}{4 \cdot 3 \cdot 2 \cdot 1} = 7 \cdot 6 \cdot 5 = 210$$

NOT ALL "RULES FOR ORDERING" ARE FUNCTIONS

We talked about the three commands "Order the cheapest one!" "Order different stuff!" and "Order what you want!" as functions on pages 37–38. It is important to note, however, that "Order different stuff!" isn't actually a function in the strictest sense, because there are several different ways to obey that command.

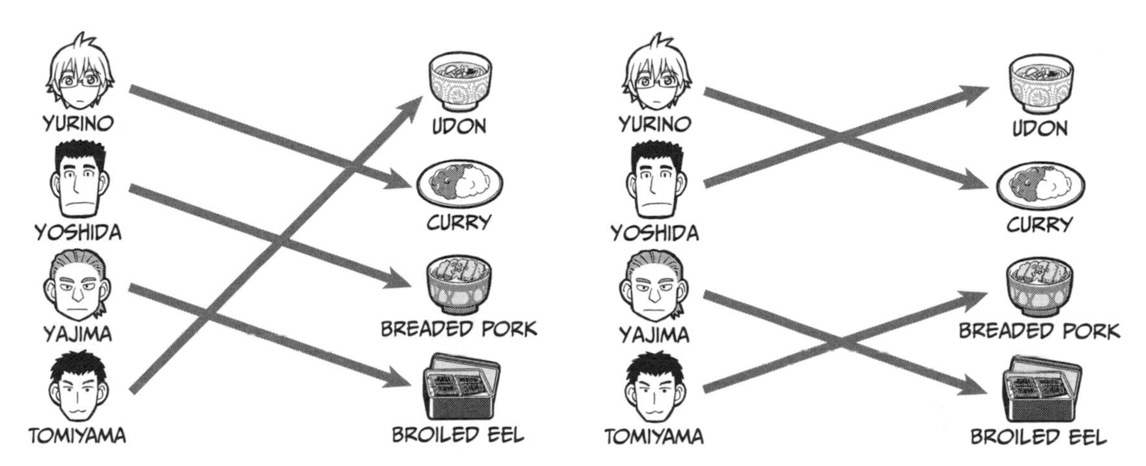

3
INTRO TO MATRICES

EI!
EI!
PUT YOUR BACKS INTO IT!
YURINO!
真派花道

DON'T RELY ON YOUR HANDS.
USE YOUR WAIST!
OSSU!
真派花道
真派花道

I THOUGHT HE'D QUIT RIGHT AWAY...

I GUESS I WAS WRONG.
HEHEH
フフン

TA-DA!
YOU MUST BE REALLY TIRED AFTER ALL THAT EXERCISE!

WOW! BUT...I COULD NEVER EAT SOMETHING SO BEAUTIFUL!
JOY
HEHE, DON'T BE SILLY.

I DON'T KNOW WHAT TO SAY...THANK YOU!

OM NOM
NOM NOM
AWESOME!
SO GOOD!
THANKS.

MISA, REALLY... THANK YOU.
DON'T WORRY ABOUT IT.

I DON'T THINK YOU'LL HAVE ANY PROBLEMS WITH THE BASICS THIS TIME AROUND EITHER.

BUT I'LL TALK A LITTLE ABOUT INVERSE MATRICES TOWARD THE END, AND THOSE CAN BE A BIT TRICKY.

OKAY.

WHAT IS A MATRIX?

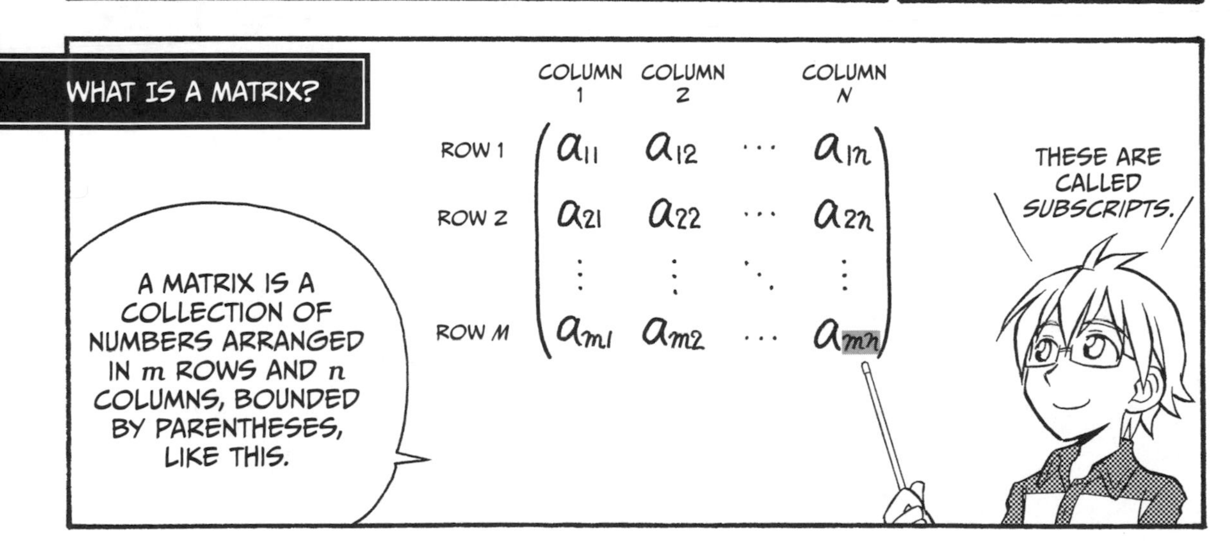

A MATRIX WITH m ROWS AND n COLUMNS IS CALLED AN "m BY n MATRIX."

$$\begin{pmatrix} 1 & 2 & 3 \\ 4 & 5 & 6 \end{pmatrix} \qquad \begin{pmatrix} -3 \\ 0 \\ 8 \\ -7 \end{pmatrix} \qquad \begin{pmatrix} a_{11} & a_{12} & \cdots & a_{1n} \\ a_{21} & a_{22} & \cdots & a_{2n} \\ \vdots & \vdots & \ddots & \vdots \\ a_{m1} & a_{m2} & \cdots & a_{mn} \end{pmatrix}$$

2×3 MATRIX 4×1 MATRIX $m \times n$ MATRIX

AH.

THE ITEMS INSIDE A MATRIX ARE CALLED ITS *ELEMENTS*.

I'VE MARKED THE (2, 1) ELEMENTS OF THESE THREE MATRICES FOR YOU.

I SEE.

	COL 1	COL 2	COL 3
ROW 1	1	2	3
ROW 2	4	5	6

	COL 1
ROW 1	−3
ROW 2	0
ROW 3	8
ROW 4	−7

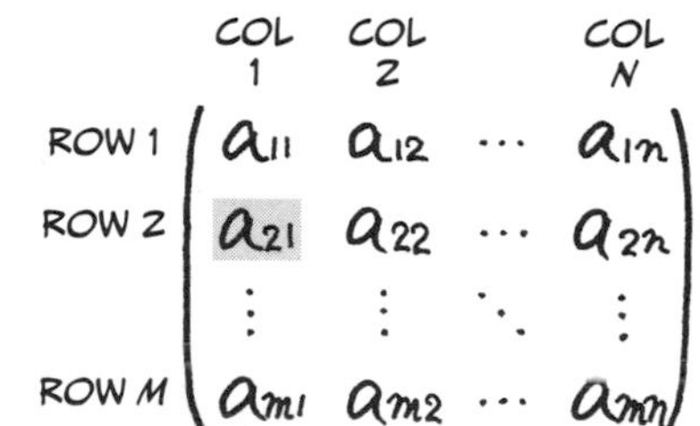

A MATRIX THAT HAS AN EQUAL NUMBER OF ROWS AND COLUMNS IS CALLED A *SQUARE MATRIX*.

$$\begin{pmatrix} 1 & 2 \\ 3 & 4 \end{pmatrix} \qquad \begin{pmatrix} a_{11} & a_{12} & \cdots & a_{1n} \\ a_{21} & a_{22} & \cdots & a_{2n} \\ \vdots & \vdots & \ddots & \vdots \\ a_{n1} & a_{n2} & \cdots & a_{nn} \end{pmatrix}$$

SQUARE MATRIX WITH TWO ROWS SQUARE MATRIX WITH n ROWS

UH HUH...

THE GRAYED OUT ELEMENTS IN THIS MATRIX ARE PART OF WHAT IS CALLED ITS *MAIN DIAGONAL*.

HMM... MATRICES AREN'T AS EXCITING AS THEY SEEM IN THE MOVIES.
UM...
YEAH. JUST NUMBERS, NO KEANU...

EXCITING OR NOT, MATRICES ARE VERY USEFUL!
WHY IS THAT?
• THEY'RE GREAT FOR WRITING LINEAR SYSTEMS MORE COMPACTLY.
• SINCE THEY MAKE FOR MORE COMPACT SYSTEMS, THEY ALSO HELP TO MAKE MATHEMATICAL LITERATURE MORE COMPACT.
• AND THEY HELP TEACHERS WRITE FASTER ON THE BLACKBOARD DURING CLASS.
WELL, THESE ARE SOME OF THE ADVANTAGES.

SO PEOPLE USE THEM BECAUSE THEY'RE PRACTICAL, HUH?
YEP.

$$\begin{cases} 1x_1 + 2x_2 = -1 \\ 3x_1 + 4x_2 = 0 \\ 5x_1 + 6x_2 = 5 \end{cases}$$

$$\begin{pmatrix} 1 & 2 \\ 3 & 4 \\ 5 & 6 \end{pmatrix}\begin{pmatrix} x_1 \\ x_2 \end{pmatrix} = \begin{pmatrix} -1 \\ 0 \\ 5 \end{pmatrix}$$

$$\begin{cases} 1x_1 + 2x_2 \\ 3x_1 + 4x_2 \\ 5x_1 + 6x_2 \end{cases}$$

$$\begin{pmatrix} 1 & 2 \\ 3 & 4 \\ 5 & 6 \end{pmatrix}\begin{pmatrix} x_1 \\ x_2 \end{pmatrix}$$

WRITING SYSTEMS OF EQUATIONS AS MATRICES

- $\begin{cases} a_{11}x_1 + a_{12}x_2 + \ldots + a_{1n}x_n = b_1 \\ a_{21}x_1 + a_{22}x_2 + \ldots + a_{2n}x_n = b_2 \\ \cdots\cdots \\ a_{m1}x_1 + a_{m2}x_2 + \ldots + a_{mn}x_n = b_m \end{cases}$ is written $\begin{pmatrix} a_{11} & a_{12} & \cdots & a_{1n} \\ a_{21} & a_{22} & \cdots & a_{2n} \\ \vdots & \vdots & \ddots & \vdots \\ a_{m1} & a_{m2} & \cdots & a_{mn} \end{pmatrix}\begin{pmatrix} x_1 \\ x_2 \\ \vdots \\ x_n \end{pmatrix} = \begin{pmatrix} b_1 \\ b_2 \\ \vdots \\ b_m \end{pmatrix}$

- $\begin{cases} a_{11}x_1 + a_{12}x_2 + \ldots + a_{1n}x_n \\ a_{21}x_1 + a_{22}x_2 + \ldots + a_{2n}x_n \\ \cdots\cdots \\ a_{m1}x_1 + a_{m2}x_2 + \ldots + a_{mn}x_n \end{cases}$ is written $\begin{pmatrix} a_{11} & a_{12} & \cdots & a_{1n} \\ a_{21} & a_{22} & \cdots & a_{2n} \\ \vdots & \vdots & \ddots & \vdots \\ a_{m1} & a_{m2} & \cdots & a_{mn} \end{pmatrix}\begin{pmatrix} x_1 \\ x_2 \\ \vdots \\ x_n \end{pmatrix}$

MATRIX CALCULATIONS

ADDITION

LET'S ADD THE 3×2 MATRIX $\begin{pmatrix} 1 & 2 \\ 3 & 4 \\ 5 & 6 \end{pmatrix}$

TO THIS 3×2 MATRIX $\begin{pmatrix} 6 & 5 \\ 4 & 3 \\ 2 & 1 \end{pmatrix}$

THAT IS: $\begin{pmatrix} 1 & 2 \\ 3 & 4 \\ 5 & 6 \end{pmatrix} + \begin{pmatrix} 6 & 5 \\ 4 & 3 \\ 2 & 1 \end{pmatrix}$

THE ELEMENTS WOULD BE ADDED ELEMENTWISE, LIKE THIS: $\begin{pmatrix} 1+6 & 2+5 \\ 3+4 & 4+3 \\ 5+2 & 6+1 \end{pmatrix}$

NOTE THAT ADDITION AND SUBTRACTION WORK ONLY WITH MATRICES THAT HAVE THE SAME DIMENSIONS.

EXAMPLES

- $\begin{pmatrix} 1 & 2 \\ 3 & 4 \\ 5 & 6 \end{pmatrix} + \begin{pmatrix} 6 & 5 \\ 4 & 3 \\ 2 & 1 \end{pmatrix} = \begin{pmatrix} 1+6 & 2+5 \\ 3+4 & 4+3 \\ 5+2 & 6+1 \end{pmatrix} = \begin{pmatrix} 7 & 7 \\ 7 & 7 \\ 7 & 7 \end{pmatrix}$

- $(10, \ 10) + (-3, \ -6) = (10 + (-3), \ 10 + (-6)) = (7, \ 4)$

- $\begin{pmatrix} 10 \\ 10 \end{pmatrix} + \begin{pmatrix} -3 \\ -6 \end{pmatrix} = \begin{pmatrix} 10 + (-3) \\ 10 + (-6) \end{pmatrix} = \begin{pmatrix} 7 \\ 4 \end{pmatrix}$

SUBTRACTION

LET'S SUBTRACT THE 3×2 MATRIX $\begin{pmatrix} 6 & 5 \\ 4 & 3 \\ 2 & 1 \end{pmatrix}$

FROM THIS 3×2 MATRIX $\begin{pmatrix} 1 & 2 \\ 3 & 4 \\ 5 & 6 \end{pmatrix}$

THAT IS: $\begin{pmatrix} 1 & 2 \\ 3 & 4 \\ 5 & 6 \end{pmatrix} - \begin{pmatrix} 6 & 5 \\ 4 & 3 \\ 2 & 1 \end{pmatrix}$

THE ELEMENTS WOULD SIMILARLY BE SUBTRACTED ELEMENTWISE, LIKE THIS: $\begin{pmatrix} 1-6 & 2-5 \\ 3-4 & 4-3 \\ 5-2 & 6-1 \end{pmatrix}$

EXAMPLES

- $\begin{pmatrix} 1 & 2 \\ 3 & 4 \\ 5 & 6 \end{pmatrix} - \begin{pmatrix} 6 & 5 \\ 4 & 3 \\ 2 & 1 \end{pmatrix} = \begin{pmatrix} 1-6 & 2-5 \\ 3-4 & 4-3 \\ 5-2 & 6-1 \end{pmatrix} = \begin{pmatrix} -5 & -3 \\ -1 & 1 \\ 3 & 5 \end{pmatrix}$

- $(10,\ 10) - (-3,\ -6) = (10 - (-3),\ 10 - (-6)) = (13,\ 16)$

- $\begin{pmatrix} 10 \\ 10 \end{pmatrix} - \begin{pmatrix} -3 \\ -6 \end{pmatrix} = \begin{pmatrix} 10 - (-3) \\ 10 - (-6) \end{pmatrix} = \begin{pmatrix} 13 \\ 16 \end{pmatrix}$

SCALAR MULTIPLICATION

LET'S MULTIPLY THE 3×2 MATRIX $\begin{pmatrix} 1 & 2 \\ 3 & 4 \\ 5 & 6 \end{pmatrix}$

BY 10. THAT IS: $10\begin{pmatrix} 1 & 2 \\ 3 & 4 \\ 5 & 6 \end{pmatrix}$

THE ELEMENTS WOULD EACH BE MULTIPLIED BY 10, LIKE THIS: $\begin{pmatrix} 10 \cdot 1 & 10 \cdot 2 \\ 10 \cdot 3 & 10 \cdot 4 \\ 10 \cdot 5 & 10 \cdot 6 \end{pmatrix}$

EXAMPLES

- $10\begin{pmatrix} 1 & 2 \\ 3 & 4 \\ 5 & 6 \end{pmatrix} = \begin{pmatrix} 10 \cdot 1 & 10 \cdot 2 \\ 10 \cdot 3 & 10 \cdot 4 \\ 10 \cdot 5 & 10 \cdot 6 \end{pmatrix} = \begin{pmatrix} 10 & 20 \\ 30 & 40 \\ 50 & 60 \end{pmatrix}$

- $2\,(3, 1) = (2 \cdot 3, 2 \cdot 1) = (6, 2)$

- $2\begin{pmatrix} 3 \\ 1 \end{pmatrix} = \begin{pmatrix} 2 \cdot 3 \\ 2 \cdot 1 \end{pmatrix} = \begin{pmatrix} 6 \\ 2 \end{pmatrix}$

MATRIX MULTIPLICATION

THE PRODUCT $\begin{pmatrix} 1 & 2 \\ 3 & 4 \\ 5 & 6 \end{pmatrix} \begin{pmatrix} x_1 & y_1 \\ x_2 & y_2 \end{pmatrix} = \begin{pmatrix} 1x_1 + 2x_2 & 1y_1 + 2y_2 \\ 3x_1 + 4x_2 & 3y_1 + 4y_2 \\ 5x_1 + 6x_2 & 5y_1 + 6y_2 \end{pmatrix}$

CAN BE DERIVED BY TEMPORARILY SEPARATING THE TWO COLUMNS $\begin{pmatrix} x_1 \\ x_2 \end{pmatrix}$ AND $\begin{pmatrix} y_1 \\ y_2 \end{pmatrix}$, FORMING THE TWO PRODUCTS

$\begin{pmatrix} 1 & 2 \\ 3 & 4 \\ 5 & 6 \end{pmatrix} \begin{pmatrix} x_1 \\ x_2 \end{pmatrix} = \begin{pmatrix} 1x_1 + 2x_2 \\ 3x_1 + 4x_2 \\ 5x_1 + 6x_2 \end{pmatrix}$ AND $\begin{pmatrix} 1 & 2 \\ 3 & 4 \\ 5 & 6 \end{pmatrix} \begin{pmatrix} y_1 \\ y_2 \end{pmatrix} = \begin{pmatrix} 1y_1 + 2y_2 \\ 3y_1 + 4y_2 \\ 5y_1 + 6y_2 \end{pmatrix}$

AND THEN REJOINING THE RESULTING COLUMNS:

$$\begin{pmatrix} 1x_1 + 2x_2 & 1y_1 + 2y_2 \\ 3x_1 + 4x_2 & 3y_1 + 4y_2 \\ 5x_1 + 6x_2 & 5y_1 + 6y_2 \end{pmatrix}$$

EXAMPLE

- $\begin{pmatrix} 1 & 2 \\ 3 & 4 \\ 5 & 6 \end{pmatrix} \begin{pmatrix} x_1 & y_1 \\ x_2 & y_2 \end{pmatrix} = \begin{pmatrix} 1x_1 + 2x_2 & 1y_1 + 2y_2 \\ 3x_1 + 4x_2 & 3y_1 + 4y_2 \\ 5x_1 + 6x_2 & 5y_1 + 6y_2 \end{pmatrix}$

- $\begin{pmatrix} 8 & -3 \\ 2 & 1 \end{pmatrix}\begin{pmatrix} 3 & 1 \\ 1 & 2 \end{pmatrix} = \begin{pmatrix} 8 \cdot 3 + (-3) \cdot 1 & 8 \cdot 1 + (-3) \cdot 2 \\ 2 \cdot 3 + 1 \cdot 1 & 2 \cdot 1 + 1 \cdot 2 \end{pmatrix} = \begin{pmatrix} 24 - 3 & 8 - 6 \\ 6 + 1 & 2 + 2 \end{pmatrix} = \begin{pmatrix} 21 & 2 \\ 7 & 4 \end{pmatrix}$

- $\begin{pmatrix} 3 & 1 \\ 1 & 2 \end{pmatrix}\begin{pmatrix} 8 & -3 \\ 2 & 1 \end{pmatrix} = \begin{pmatrix} 3 \cdot 8 + 1 \cdot 2 & 3 \cdot (-3) + 1 \cdot 1 \\ 1 \cdot 8 + 2 \cdot 2 & 1 \cdot (-3) + 2 \cdot 1 \end{pmatrix} = \begin{pmatrix} 24 + 2 & -9 + 1 \\ 8 + 4 & -3 + 2 \end{pmatrix} = \begin{pmatrix} 26 & -8 \\ 12 & -1 \end{pmatrix}$

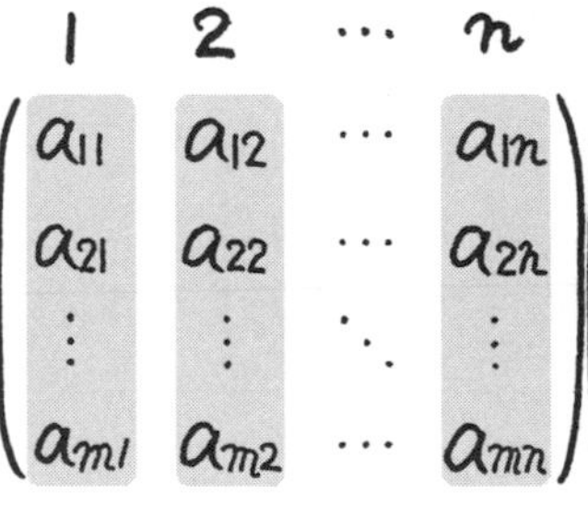

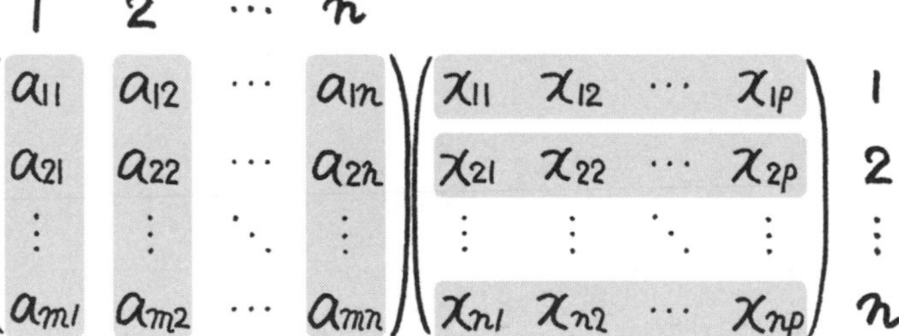

$$\begin{matrix} 1 & 2 & \cdots & n \end{matrix}$$
$$\begin{pmatrix} a_{11} & a_{12} & \cdots & a_{1n} \\ a_{21} & a_{22} & \cdots & a_{2n} \\ \vdots & \vdots & \ddots & \vdots \\ a_{m1} & a_{m2} & \cdots & a_{mn} \end{pmatrix}\begin{pmatrix} x_{11} & x_{12} & \cdots & x_{1p} \\ x_{21} & x_{22} & \cdots & x_{2p} \\ \vdots & \vdots & \ddots & \vdots \\ x_{n1} & x_{n2} & \cdots & x_{np} \end{pmatrix} \begin{matrix} 1 \\ 2 \\ \vdots \\ n \end{matrix}$$

AN $m \times n$ MATRIX TIMES AN $n \times p$ MATRIX YIELDS AN $m \times p$ MATRIX.

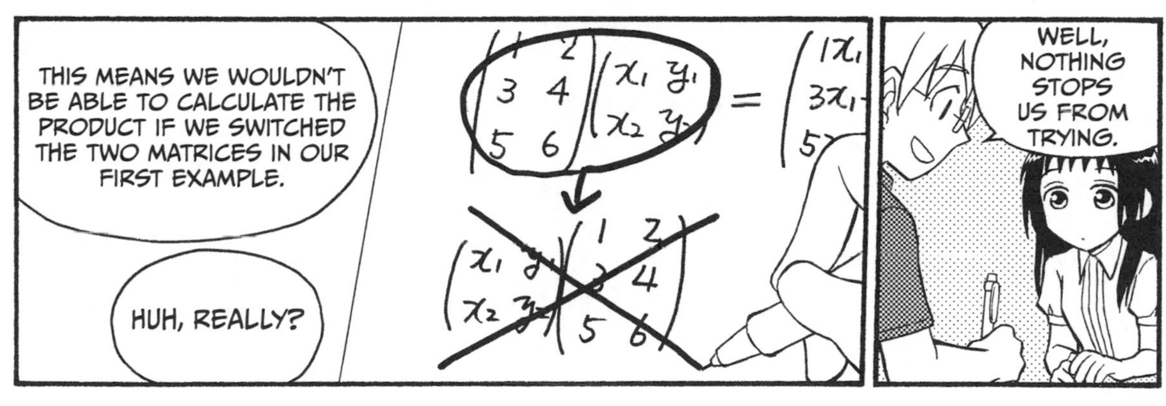

PRODUCT OF 3×2 AND 2×2 FACTORS	$\begin{pmatrix}1&2\\3&4\\5&6\end{pmatrix}\begin{pmatrix}x_1&y_1\\x_2&y_2\end{pmatrix}$ IS THE SAME AS $\begin{pmatrix}1&2\\3&4\\5&6\end{pmatrix}\begin{pmatrix}x_1\\x_2\end{pmatrix}$ AND $\begin{pmatrix}1&2\\3&4\\5&6\end{pmatrix}\begin{pmatrix}y_1\\y_2\end{pmatrix}$ WHICH IS THE SAME AS $\begin{cases}1x_1+2x_2\\3x_1+4x_2\\5x_1+6x_2\end{cases}$ $\begin{cases}1y_1+2y_2\\3y_1+4y_2\\5y_1+6y_2\end{cases}$ IN THE SAME MATRIX.
PRODUCT OF 2×2 AND 3×2 FACTORS	$\begin{pmatrix}x_1&y_1\\x_2&y_2\end{pmatrix}\begin{pmatrix}1&2\\3&4\\5&6\end{pmatrix}$ IS THE SAME AS $\begin{pmatrix}x_1&y_1\\x_2&y_2\end{pmatrix}\begin{pmatrix}1\\3\\5\end{pmatrix}$ AND $\begin{pmatrix}x_1&y_1\\x_2&y_2\end{pmatrix}\begin{pmatrix}2\\4\\6\end{pmatrix}$ WHICH IS THE SAME AS $\begin{cases}x_1\cdot 1+y_1\cdot 3+?\cdot 5\\x_2\cdot 1+y_2\cdot 3+?\cdot 5\end{cases}$ AND $\begin{cases}x_1\cdot 2+y_1\cdot 4+?\cdot 6\\x_2\cdot 2+y_2\cdot 4+?\cdot 6\end{cases}$ IN THE SAME MATRIX.

WE RUN INTO A PROBLEM HERE: THERE ARE NO ELEMENTS CORRESPONDING TO THESE POSITIONS!

OOPS...

ONE MORE THING. IT'S OKAY TO USE EXPONENTS TO EXPRESS REPEATED MULTIPLICATION OF SQUARE MATRICES.

$$\underbrace{\begin{pmatrix}a_{11}&a_{12}&\cdots&a_{1n}\\a_{21}&a_{22}&\cdots&a_{2n}\\\vdots&\vdots&\ddots&\vdots\\a_{n1}&a_{n2}&\cdots&a_{nn}\end{pmatrix}\begin{pmatrix}a_{11}&a_{12}&\cdots&a_{1n}\\a_{21}&a_{22}&\cdots&a_{2n}\\\vdots&\vdots&\ddots&\vdots\\a_{n1}&a_{n2}&\cdots&a_{nn}\end{pmatrix}\cdots\begin{pmatrix}a_{11}&a_{12}&\cdots&a_{1n}\\a_{21}&a_{22}&\cdots&a_{2n}\\\vdots&\vdots&\ddots&\vdots\\a_{n1}&a_{n2}&\cdots&a_{nn}\end{pmatrix}}_{p \text{ FACTORS}}$$

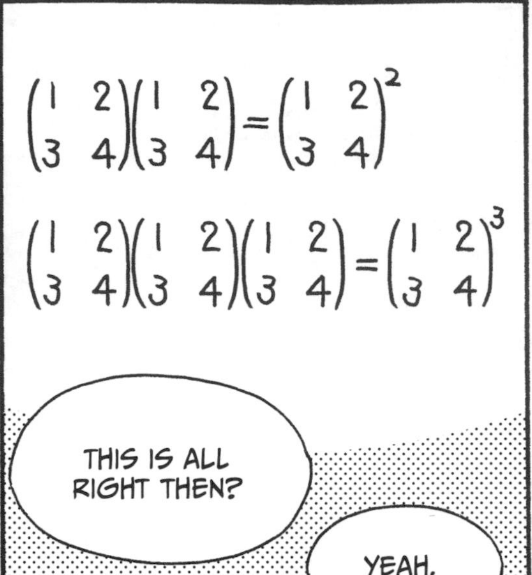

THE EASIEST WAY WOULD BE TO JUST MULTIPLY THEM FROM LEFT TO RIGHT, LIKE THIS:

$$\begin{pmatrix} 1 & 2 \\ 3 & 4 \end{pmatrix}^3 = \begin{pmatrix} 1 & 2 \\ 3 & 4 \end{pmatrix}\begin{pmatrix} 1 & 2 \\ 3 & 4 \end{pmatrix}\begin{pmatrix} 1 & 2 \\ 3 & 4 \end{pmatrix} = \begin{pmatrix} 1\cdot 1 + 2\cdot 3 & 1\cdot 2 + 2\cdot 4 \\ 3\cdot 1 + 4\cdot 3 & 3\cdot 2 + 4\cdot 4 \end{pmatrix}\begin{pmatrix} 1 & 2 \\ 3 & 4 \end{pmatrix}$$

$$\begin{pmatrix} 7 & 10 \\ 15 & 22 \end{pmatrix}\begin{pmatrix} 1 & 2 \\ 3 & 4 \end{pmatrix} = \begin{pmatrix} 7\cdot 1 + 10\cdot 3 & 7\cdot 2 + 10\cdot 4 \\ 15\cdot 1 + 22\cdot 3 & 15\cdot 2 + 22\cdot 4 \end{pmatrix} = \begin{pmatrix} 37 & 54 \\ 81 & 118 \end{pmatrix}$$

SO WE'LL LOOK AT ONLY THESE EIGHT TODAY.

1. ZERO MATRICES
2. TRANSPOSE MATRICES
3. SYMMETRIC MATRICES
4. UPPER TRIANGULAR MATRICES
5. LOWER TRIANGULAR MATRICES
6. DIAGONAL MATRICES
7. IDENTITY MATRICES
8. INVERSE MATRICES

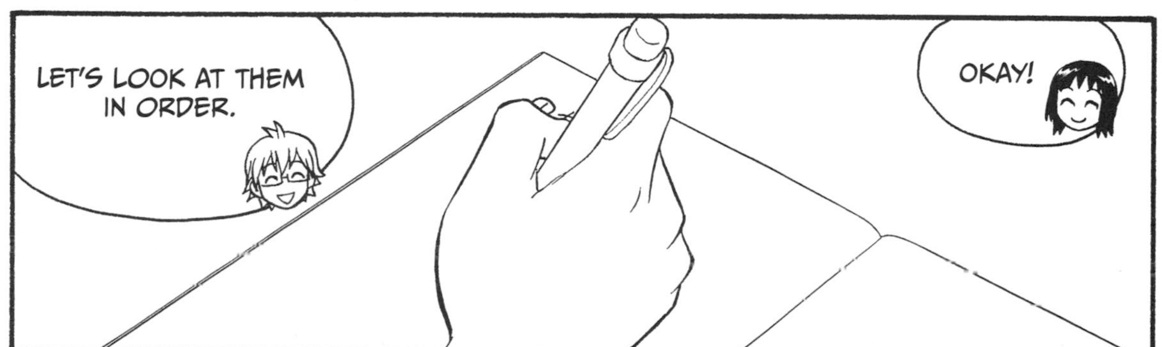

❶ ZERO MATRICES

A zero matrix is a matrix where all elements are equal to zero.

$$\begin{pmatrix} 0 & 0 \\ 0 & 0 \end{pmatrix} \qquad \begin{pmatrix} 0 & 0 & 0 \\ 0 & 0 & 0 \end{pmatrix} \qquad \begin{pmatrix} 0 \\ 0 \\ 0 \\ 0 \end{pmatrix}$$

❷ TRANSPOSE MATRICES

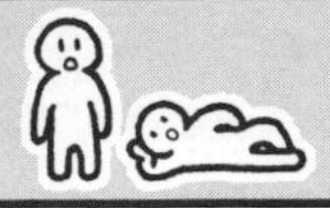

The easiest way to understand transpose matrices is to just look at an example.

If we transpose the 2×3 matrix $\begin{pmatrix} 1 & 3 & 5 \\ 2 & 4 & 6 \end{pmatrix}$

we get the 3×2 matrix $\begin{pmatrix} 1 & 2 \\ 3 & 4 \\ 5 & 6 \end{pmatrix}$

As you can see, the transpose operator switches the rows and columns in a matrix.

The transpose of the $n \times m$ matrix $\begin{pmatrix} a_{11} & a_{21} & \cdots & a_{m1} \\ a_{12} & a_{22} & \cdots & a_{m2} \\ \vdots & \vdots & \ddots & \vdots \\ a_{1n} & a_{2n} & \cdots & a_{mn} \end{pmatrix}$

is consequently $\begin{pmatrix} a_{11} & a_{12} & \cdots & a_{1n} \\ a_{21} & a_{22} & \cdots & a_{2n} \\ \vdots & \vdots & \ddots & \vdots \\ a_{m1} & a_{m2} & \cdots & a_{mn} \end{pmatrix}$

The most common way to indicate a transpose is to add a small *T* at the top-right corner of the matrix.

$$\begin{pmatrix} a_{11} & a_{12} & \cdots & a_{1n} \\ a_{21} & a_{22} & \cdots & a_{2n} \\ \vdots & \vdots & \ddots & \vdots \\ a_{m1} & a_{m2} & \cdots & a_{mn} \end{pmatrix}^{T}$$

For example:

$$\begin{pmatrix} 1 & 3 & 5 \\ 2 & 4 & 6 \end{pmatrix}^{T} = \begin{pmatrix} 1 & 2 \\ 3 & 4 \\ 5 & 6 \end{pmatrix}$$

❸ SYMMETRIC MATRICES

Symmetric matrices are square matrices that are symmetric around their main diagonals.

$$\begin{pmatrix} 1 & 5 & 6 & 7 \\ 5 & 2 & 8 & 9 \\ 6 & 8 & 3 & 10 \\ 7 & 9 & 10 & 4 \end{pmatrix}$$

Because of this characteristic, a symmetric matrix is always equal to its transpose.

❹ UPPER TRIANGULAR AND ❺ LOWER TRIANGULAR MATRICES

Triangular matrices are square matrices in which the elements either above the main diagonal or below it are all equal to zero.

This is an upper triangular matrix, since all elements *below* the main diagonal are zero.

$$\begin{pmatrix} 1 & 5 & 6 & 7 \\ 0 & 2 & 8 & 9 \\ 0 & 0 & 3 & 10 \\ 0 & 0 & 0 & 4 \end{pmatrix}$$

This is a lower triangular matrix—all elements *above* the main diagonal are zero.

$$\begin{pmatrix} 1 & 0 & 0 & 0 \\ 5 & 2 & 0 & 0 \\ 6 & 8 & 3 & 0 \\ 7 & 9 & 10 & 4 \end{pmatrix}$$

❻ DIAGONAL MATRICES

A diagonal matrix is a square matrix in which all elements that are not part of its main diagonal are equal to zero.

For example, $\begin{pmatrix} 1 & 0 & 0 & 0 \\ 0 & 2 & 0 & 0 \\ 0 & 0 & 3 & 0 \\ 0 & 0 & 0 & 4 \end{pmatrix}$ is a diagonal matrix.

Note that this matrix could also be written as diag(1,2,3,4).

SEE FOR YOURSELF!

$$\begin{pmatrix} a_{11} & 0 & \cdots & 0 \\ 0 & a_{22} & \cdots & 0 \\ \vdots & \vdots & \ddots & \vdots \\ 0 & 0 & \cdots & a_{nn} \end{pmatrix}^p = \begin{pmatrix} a_{11}^p & 0 & \cdots & 0 \\ 0 & a_{22}^p & \cdots & 0 \\ \vdots & \vdots & \ddots & \vdots \\ 0 & 0 & \cdots & a_{nn}^p \end{pmatrix}$$

TRY CALCULATING

$\begin{pmatrix} 2 & 0 \\ 0 & 3 \end{pmatrix}^2$ AND $\begin{pmatrix} 2 & 0 \\ 0 & 3 \end{pmatrix}^3$

TO SEE WHY.

HMM...

$$\cdot \begin{pmatrix} 2 & 0 \\ 0 & 3 \end{pmatrix}^2 = \begin{pmatrix} 2 & 0 \\ 0 & 3 \end{pmatrix}\begin{pmatrix} 2 & 0 \\ 0 & 3 \end{pmatrix} = \begin{pmatrix} 2 \cdot 2 + 0 \cdot 0 & 2 \cdot 0 + 0 \cdot 3 \\ 0 \cdot 2 + 3 \cdot 0 & 0 \cdot 0 + 3 \cdot 3 \end{pmatrix} = \begin{pmatrix} 2^2 & 0 \\ 0 & 3^2 \end{pmatrix}$$

$$\cdot \begin{pmatrix} 2 & 0 \\ 0 & 3 \end{pmatrix}^3 = \begin{pmatrix} 2 & 0 \\ 0 & 3 \end{pmatrix}^2\begin{pmatrix} 2 & 0 \\ 0 & 3 \end{pmatrix} = \begin{pmatrix} 2^2 & 0 \\ 0 & 3^2 \end{pmatrix}\begin{pmatrix} 2 & 0 \\ 0 & 3 \end{pmatrix} = \begin{pmatrix} 2^2 \cdot 2 + 0 \cdot 0 & 2^2 \cdot 0 + 0 \cdot 3 \\ 0 \cdot 2 + 3^2 \cdot 0 & 0 \cdot 0 + 3^2 \cdot 3 \end{pmatrix} = \begin{pmatrix} 2^3 & 0 \\ 0 & 3^3 \end{pmatrix}$$

LIKE THIS?

❼ IDENTITY MATRICES

Identity matrices are in essence diag(1,1,1,...,1). In other words, they are square matrices with n rows in which all elements on the main diagonal are equal to 1 and all other elements are 0.

For example, an identity matrix with $n = 4$ would look like this:

$$\begin{pmatrix} 1 & 0 & 0 & 0 \\ 0 & 1 & 0 & 0 \\ 0 & 0 & 1 & 0 \\ 0 & 0 & 0 & 1 \end{pmatrix}$$

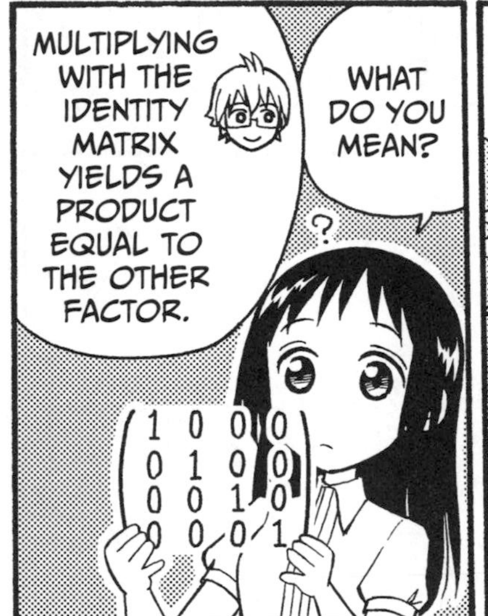

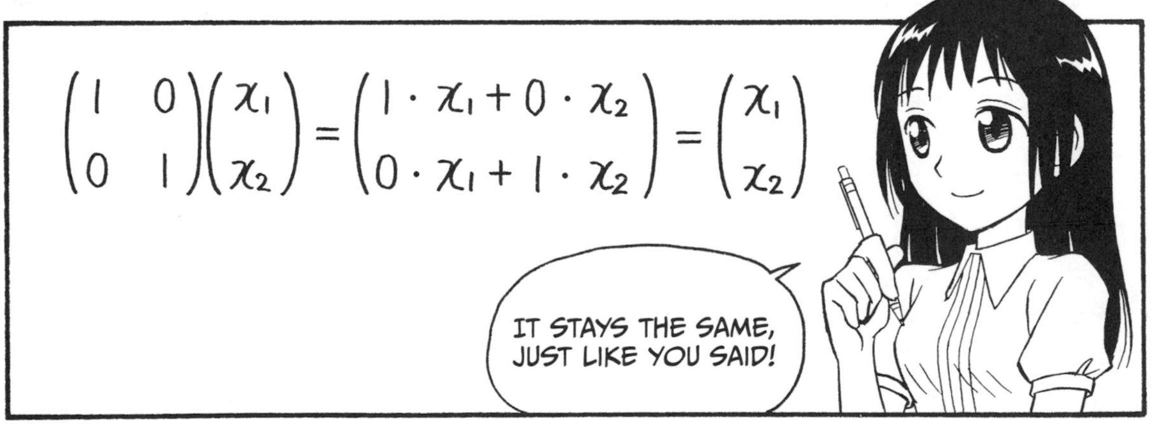

LET'S TRY A FEW OTHER EXAMPLES.

- $$\begin{pmatrix} 1 & 0 & \cdots & 0 \\ 0 & 1 & \cdots & 0 \\ \vdots & \vdots & \ddots & \vdots \\ 0 & 0 & \cdots & 1 \end{pmatrix} \begin{pmatrix} x_1 \\ x_2 \\ \vdots \\ x_n \end{pmatrix} = \begin{pmatrix} 1 \cdot x_1 + 0 \cdot x_2 + \cdots + 0 \cdot x_n \\ 0 \cdot x_1 + 1 \cdot x_2 + \cdots + 0 \cdot x_n \\ \vdots \\ 0 \cdot x_1 + 0 \cdot x_2 + \cdots + 1 \cdot x_n \end{pmatrix} = \begin{pmatrix} x_1 \\ x_2 \\ \vdots \\ x_n \end{pmatrix}$$

- $$\begin{pmatrix} 1 & 0 \\ 0 & 1 \end{pmatrix} \begin{pmatrix} x_{11} & x_{21} & \cdots & x_{n1} \\ x_{12} & x_{22} & \cdots & x_{n2} \end{pmatrix} = \begin{pmatrix} 1 \cdot x_{11} + 0 \cdot x_{12} & 1 \cdot x_{21} + 0 \cdot x_{22} & \cdots & 1 \cdot x_{n1} + 0 \cdot x_{n2} \\ 0 \cdot x_{11} + 1 \cdot x_{12} & 0 \cdot x_{21} + 1 \cdot x_{22} & \cdots & 0 \cdot x_{n1} + 1 \cdot x_{n2} \end{pmatrix}$$
$$= \begin{pmatrix} x_{11} & x_{21} & \cdots & x_{n1} \\ x_{12} & x_{22} & \cdots & x_{n2} \end{pmatrix}$$

- $$\begin{pmatrix} x_{11} & x_{12} \\ x_{21} & x_{22} \\ \vdots & \vdots \\ x_{n1} & x_{n2} \end{pmatrix} \begin{pmatrix} 1 & 0 \\ 0 & 1 \end{pmatrix} = \begin{pmatrix} x_{11} \cdot 1 + x_{12} \cdot 0 & x_{11} \cdot 0 + x_{12} \cdot 1 \\ x_{21} \cdot 1 + x_{22} \cdot 0 & x_{21} \cdot 0 + x_{22} \cdot 1 \\ \vdots & \vdots \\ x_{n1} \cdot 1 + x_{n2} \cdot 0 & x_{n1} \cdot 0 + x_{n2} \cdot 1 \end{pmatrix} = \begin{pmatrix} x_{11} & x_{12} \\ x_{21} & x_{22} \\ \vdots & \vdots \\ x_{n1} & x_{n2} \end{pmatrix}$$

LET'S TAKE A BREAK. WE STILL HAVE INVERSE MATRICES LEFT TO LOOK AT, BUT THEY'RE A BIT MORE COMPLEX THAN THIS.
FINE BY ME.
AH!

THANKS AGAIN FOR LUNCH. I HAD NO IDEA YOU WERE SUCH A GOOD COOK.

NO PROBLEM!

I'LL MAKE YOU ANOTHER TOMORROW IF YOU'D LIKE.
NO, I WASN'T TRYING TO GET YOU TO—

DON'T WORRY ABOUT IT. THE BEST PART OF COOKING SOMETHING IS SEEING SOMEONE ENJOY IT.
IT'D BE A PLEASURE.

TH-THANKS...

WHAT DO YOU SAY? READY FOR MORE MATRICES?
SURE.

4
MORE MATRICES

❽ INVERSE MATRICES

If the product of two square matrices is an identity matrix, then the two factor matrices are *inverses* of each other.

This means that $\begin{pmatrix} x_{11} & x_{12} \\ x_{21} & x_{22} \end{pmatrix}$ is an inverse matrix to $\begin{pmatrix} 1 & 2 \\ 3 & 4 \end{pmatrix}$ if

$$\begin{pmatrix} 1 & 2 \\ 3 & 4 \end{pmatrix}\begin{pmatrix} x_{11} & x_{12} \\ x_{21} & x_{22} \end{pmatrix} = \begin{pmatrix} 1 & 0 \\ 0 & 1 \end{pmatrix}$$

AND THAT'S IT.
HUH?
DIDN'T YOU SAY SOMETHING ABOUT EXAMPLES? WE'RE DONE ALREADY?
DON'T WORRY, THAT WAS ONLY THE DEFINITION!

SINCE THEY'RE SO IMPORTANT, I THOUGHT WE'D GO INTO MORE DETAIL ON THIS ONE.
I'LL TEACH YOU HOW TO IDENTIFY WHETHER AN INVERSE EXISTS OR NOT—AND ALSO HOW TO CALCULATE ONE.
$\begin{pmatrix} 1 & 2 \\ 3 & 4 \end{pmatrix}\begin{pmatrix} x_{11} & x_{12} \\ x_{21} & x_{22} \end{pmatrix} = \begin{pmatrix} 1 & 0 \\ 0 & 1 \end{pmatrix}$

SHOULD WE GET RIGHT DOWN TO BUSINESS?
SURE!

CALCULATING INVERSE MATRICES

* THE JAPANESE TERM FOR GAUSSIAN ELIMINATION IS *HAKIDASHIHOU*, WHICH ROUGHLY TRANSLATES TO "THE SWEEPING OUT METHOD." KEEP THIS IN MIND AS YOU'RE READING THIS CHAPTER!

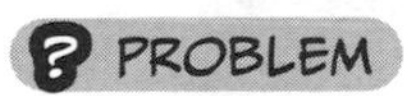

Solve the following linear system:

$$\begin{cases} 3x_1 + 1x_2 = 1 \\ 1x_1 + 2x_2 = 0 \end{cases}$$

SOLUTION

KEEP COMPARING THE ROWS ON THE LEFT TO SEE HOW IT WORKS.

OKAY.

THE COMMON METHOD	THE COMMON METHOD EXPRESSED WITH MATRICES	GAUSSIAN ELIMINATION
$\begin{cases} 3x_1 + 1x_2 = 1 \\ 1x_1 + 2x_2 = 0 \end{cases}$ Start by multiplying the top equation by 2.	$\begin{pmatrix} 3 & 1 \\ 1 & 2 \end{pmatrix}\begin{pmatrix} x_1 \\ x_2 \end{pmatrix} = \begin{pmatrix} 1 \\ 0 \end{pmatrix}$	$\begin{pmatrix} 3 & 1 & 1 \\ 1 & 2 & 0 \end{pmatrix}$
$\begin{cases} 6x_1 + 2x_2 = 2 \\ 1x_1 + 2x_2 = 0 \end{cases}$ Subtract the bottom equation from the top equation.	$\begin{pmatrix} 6 & 2 \\ 1 & 2 \end{pmatrix}\begin{pmatrix} x_1 \\ x_2 \end{pmatrix} = \begin{pmatrix} 2 \\ 0 \end{pmatrix}$	$\begin{pmatrix} 6 & 2 & 2 \\ 1 & 2 & 0 \end{pmatrix}$
$\begin{cases} 5x_1 + 0x_2 = 2 \\ 1x_1 + 2x_2 = 0 \end{cases}$ Multiply the bottom equation by 5.	$\begin{pmatrix} 5 & 0 \\ 1 & 2 \end{pmatrix}\begin{pmatrix} x_1 \\ x_2 \end{pmatrix} = \begin{pmatrix} 2 \\ 0 \end{pmatrix}$	$\begin{pmatrix} 5 & 0 & 2 \\ 1 & 2 & 0 \end{pmatrix}$
$\begin{cases} 5x_1 + 0x_2 = 2 \\ 5x_1 + 10x_2 = 0 \end{cases}$ Subtract the top equation from the bottom equation.	$\begin{pmatrix} 5 & 0 \\ 5 & 10 \end{pmatrix}\begin{pmatrix} x_1 \\ x_2 \end{pmatrix} = \begin{pmatrix} 2 \\ 0 \end{pmatrix}$	$\begin{pmatrix} 5 & 0 & 2 \\ 5 & 10 & 0 \end{pmatrix}$
$\begin{cases} 5x_1 + 0x_2 = 2 \\ 0x_1 + 10x_2 = -2 \end{cases}$ Divide the top equation by 5 and the bottom by 10.	$\begin{pmatrix} 5 & 0 \\ 0 & 10 \end{pmatrix}\begin{pmatrix} x_1 \\ x_2 \end{pmatrix} = \begin{pmatrix} 2 \\ -2 \end{pmatrix}$	$\begin{pmatrix} 5 & 0 & 2 \\ 0 & 10 & -2 \end{pmatrix}$
$\begin{cases} 1x_1 + 0x_2 = \frac{2}{5} \\ 0x_1 + 1x_2 = -\frac{1}{5} \end{cases}$ And we're done!	$\begin{pmatrix} 1 & 0 \\ 0 & 1 \end{pmatrix}\begin{pmatrix} x_1 \\ x_2 \end{pmatrix} = \begin{pmatrix} \frac{2}{5} \\ -\frac{1}{5} \end{pmatrix}$	$\begin{pmatrix} 1 & 0 & \frac{2}{5} \\ 0 & 1 & -\frac{1}{5} \end{pmatrix}$

GATHER 'EM UP AND SWEEP.

GATHER 'EM UP AND SWEEP.

DONE!

PROBLEM

Find the inverse of the 2×2 matrix $\begin{pmatrix} 3 & 1 \\ 1 & 2 \end{pmatrix}$

We're trying to find the inverse of $\begin{pmatrix} 3 & 1 \\ 1 & 2 \end{pmatrix}$

↓

We need to find the matrix $\begin{pmatrix} x_{11} & x_{12} \\ x_{21} & x_{22} \end{pmatrix}$ that satisfies $\begin{pmatrix} 3 & 1 \\ 1 & 2 \end{pmatrix}\begin{pmatrix} x_{11} & x_{12} \\ x_{21} & x_{22} \end{pmatrix} = \begin{pmatrix} 1 & 0 \\ 0 & 1 \end{pmatrix}$

↓

or $\begin{pmatrix} x_{11} \\ x_{21} \end{pmatrix}$ and $\begin{pmatrix} x_{12} \\ x_{22} \end{pmatrix}$ that satisfy $\begin{cases} \begin{pmatrix} 3 & 1 \\ 1 & 2 \end{pmatrix}\begin{pmatrix} x_{11} \\ x_{21} \end{pmatrix} = \begin{pmatrix} 1 \\ 0 \end{pmatrix} \\ \begin{pmatrix} 3 & 1 \\ 1 & 2 \end{pmatrix}\begin{pmatrix} x_{12} \\ x_{22} \end{pmatrix} = \begin{pmatrix} 0 \\ 1 \end{pmatrix} \end{cases}$

↓

We need to solve the systems $\begin{cases} 3x_{11} + 1x_{21} = 1 \\ 1x_{11} + 2x_{21} = 0 \end{cases}$ and $\begin{cases} 3x_{12} + 1x_{22} = 0 \\ 1x_{12} + 2x_{22} = 1 \end{cases}$

SOLUTION

THE COMMON METHOD	THE COMMON METHOD EXPRESSED WITH MATRICES	GAUSSIAN ELIMINATION
$\begin{cases}3x_{11}+1x_{21}=1\\1x_{11}+2x_{21}=0\end{cases}$ $\begin{cases}3x_{12}+1x_{22}=0\\1x_{12}+2x_{22}=1\end{cases}$ Multiply the top equation by 2.	$\begin{pmatrix}3&1\\1&2\end{pmatrix}\begin{pmatrix}x_{11}&x_{12}\\x_{21}&x_{22}\end{pmatrix}=\begin{pmatrix}1&0\\0&1\end{pmatrix}$	$\begin{pmatrix}3&1&1&0\\1&2&0&1\end{pmatrix}$
$\begin{cases}6x_{11}+2x_{21}=2\\1x_{11}+2x_{21}=0\end{cases}$ $\begin{cases}6x_{12}+2x_{22}=0\\1x_{12}+2x_{22}=1\end{cases}$ Subtract the bottom equation from the top.	$\begin{pmatrix}6&2\\1&2\end{pmatrix}\begin{pmatrix}x_{11}&x_{12}\\x_{21}&x_{22}\end{pmatrix}=\begin{pmatrix}2&0\\0&1\end{pmatrix}$	$\begin{pmatrix}6&2&2&0\\1&2&0&1\end{pmatrix}$
$\begin{cases}5x_{11}+0x_{21}=2\\1x_{11}+2x_{21}=0\end{cases}$ $\begin{cases}5x_{12}+0x_{22}=-1\\1x_{12}+2x_{22}=1\end{cases}$ Multiply the bottom equation by 5.	$\begin{pmatrix}5&0\\1&2\end{pmatrix}\begin{pmatrix}x_{11}&x_{12}\\x_{21}&x_{22}\end{pmatrix}=\begin{pmatrix}2&-1\\0&1\end{pmatrix}$	$\begin{pmatrix}5&0&2&-1\\1&2&0&1\end{pmatrix}$
$\begin{cases}5x_{11}+0x_{21}=2\\5x_{11}+10x_{21}=0\end{cases}$ $\begin{cases}5x_{12}+0x_{22}=-1\\5x_{12}+10x_{22}=5\end{cases}$ Subtract the top equation from the bottom.	$\begin{pmatrix}5&0\\5&10\end{pmatrix}\begin{pmatrix}x_{11}&x_{12}\\x_{21}&x_{22}\end{pmatrix}=\begin{pmatrix}2&-1\\0&5\end{pmatrix}$	$\begin{pmatrix}5&0&2&-1\\5&10&0&5\end{pmatrix}$
$\begin{cases}5x_{11}+0x_{21}=2\\0x_{11}+10x_{21}=-2\end{cases}$ $\begin{cases}5x_{12}+0x_{22}=-1\\0x_{12}+10x_{22}=6\end{cases}$ Divide the top by 5 and the bottom by 10.	$\begin{pmatrix}5&0\\0&10\end{pmatrix}\begin{pmatrix}x_{11}&x_{12}\\x_{21}&x_{22}\end{pmatrix}=\begin{pmatrix}2&-1\\-2&6\end{pmatrix}$	$\begin{pmatrix}5&0&2&-1\\0&10&-2&6\end{pmatrix}$
$\begin{cases}1x_{11}+0x_{21}=\frac{2}{5}\\0x_{11}+1x_{21}=-\frac{1}{5}\end{cases}$ $\begin{cases}1x_{12}+0x_{22}=-\frac{1}{5}\\0x_{12}+1x_{22}=\frac{3}{5}\end{cases}$ This is our inverse matrix; we're done!	$\begin{pmatrix}1&0\\0&1\end{pmatrix}\begin{pmatrix}x_{11}&x_{12}\\x_{21}&x_{22}\end{pmatrix}=\begin{pmatrix}\frac{2}{5}&-\frac{1}{5}\\-\frac{1}{5}&\frac{3}{5}\end{pmatrix}$	$\begin{pmatrix}1&0&\frac{2}{5}&-\frac{1}{5}\\0&1&-\frac{1}{5}&\frac{3}{5}\end{pmatrix}$

LET'S MAKE SURE THAT THE PRODUCT OF THE ORIGINAL AND CALCULATED MATRICES REALLY IS THE IDENTITY MATRIX.

The product of the original and inverse matrix is

- $$\begin{pmatrix} 3 & 1 \\ 1 & 2 \end{pmatrix}\begin{pmatrix} \frac{2}{5} & -\frac{1}{5} \\ -\frac{1}{5} & \frac{3}{5} \end{pmatrix} = \begin{pmatrix} 3 \cdot \frac{2}{5} + 1 \cdot \left(-\frac{1}{5}\right) & 3 \cdot \left(-\frac{1}{5}\right) + 1 \cdot \frac{3}{5} \\ 1 \cdot \frac{2}{5} + 2 \cdot \left(-\frac{1}{5}\right) & 1 \cdot \left(-\frac{1}{5}\right) + 2 \cdot \frac{3}{5} \end{pmatrix} = \begin{pmatrix} 1 & 0 \\ 0 & 1 \end{pmatrix}$$

The product of the inverse and original matrix is

- $$\begin{pmatrix} \frac{2}{5} & -\frac{1}{5} \\ -\frac{1}{5} & \frac{3}{5} \end{pmatrix}\begin{pmatrix} 3 & 1 \\ 1 & 2 \end{pmatrix} = \begin{pmatrix} \frac{2}{5} \cdot 3 + \left(-\frac{1}{5}\right) \cdot 1 & \frac{2}{5} \cdot 1 + \left(-\frac{1}{5}\right) \cdot 2 \\ \left(-\frac{1}{5}\right) \cdot 3 + \frac{3}{5} \cdot 1 & \left(-\frac{1}{5}\right) \cdot 1 + \frac{3}{5} \cdot 2 \end{pmatrix} = \begin{pmatrix} 1 & 0 \\ 0 & 1 \end{pmatrix}$$

IT SEEMS LIKE THEY BOTH BECOME THE IDENTITY MATRIX...

THAT'S AN IMPORTANT POINT: THE ORDER OF THE FACTORS DOESN'T MATTER. THE PRODUCT IS ALWAYS THE IDENTITY MATRIX! REMEMBERING THIS TEST IS VERY USEFUL. YOU SHOULD USE IT AS OFTEN AS YOU CAN TO CHECK YOUR CALCULATIONS.

BY THE WAY...

THE SYMBOL USED TO DENOTE INVERSE MATRICES IS THE SAME AS ANY INVERSE IN MATHEMATICS, SO...

THE INVERSE OF

$$\begin{pmatrix} a_{11} & a_{12} & \cdots & a_{1n} \\ a_{21} & a_{22} & \cdots & a_{2n} \\ \vdots & \vdots & \ddots & \vdots \\ a_{n1} & a_{n2} & \cdots & a_{nn} \end{pmatrix}$$

IS WRITTEN AS

$$\begin{pmatrix} a_{11} & a_{12} & \cdots & a_{1n} \\ a_{21} & a_{22} & \cdots & a_{2n} \\ \vdots & \vdots & \ddots & \vdots \\ a_{n1} & a_{n2} & \cdots & a_{nn} \end{pmatrix}^{-1}$$

ACTUALLY...WE ALSO COULD HAVE SOLVED $\begin{pmatrix} a_{11} & a_{12} \\ a_{21} & a_{22} \end{pmatrix}^{-1}$ WITH...

$\begin{pmatrix} a_{11} & a_{12} \\ a_{21} & a_{22} \end{pmatrix}^{-1} = \frac{1}{a_{11}a_{22} - a_{12}a_{21}} \begin{pmatrix} a_{22} & -a_{12} \\ -a_{21} & a_{11} \end{pmatrix}$
...THIS FORMULA RIGHT HERE.
HUH?

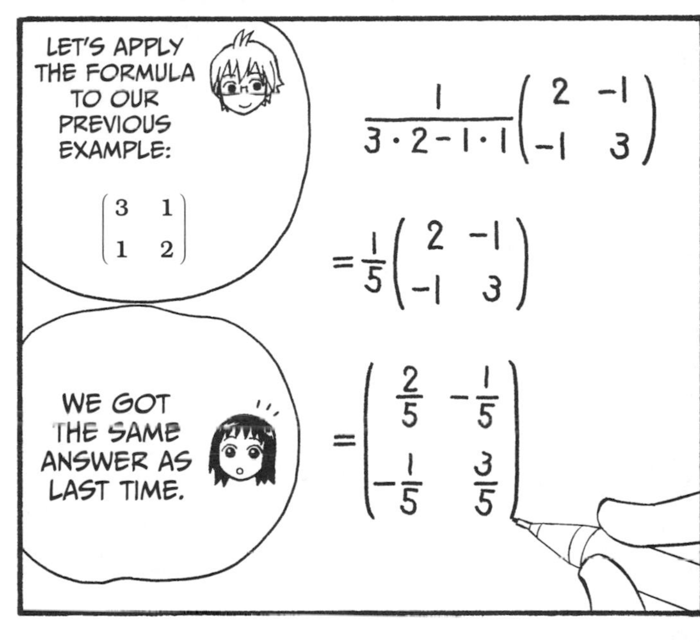
LET'S APPLY THE FORMULA TO OUR PREVIOUS EXAMPLE:
$\begin{pmatrix} 3 & 1 \\ 1 & 2 \end{pmatrix}$
$\frac{1}{3 \cdot 2 - 1 \cdot 1} \begin{pmatrix} 2 & -1 \\ -1 & 3 \end{pmatrix} = \frac{1}{5} \begin{pmatrix} 2 & -1 \\ -1 & 3 \end{pmatrix} = \begin{pmatrix} \frac{2}{5} & -\frac{1}{5} \\ -\frac{1}{5} & \frac{3}{5} \end{pmatrix}$
WE GOT THE SAME ANSWER AS LAST TIME.

WHY EVEN BOTHER WITH THE OTHER METHOD?
AH, WELL...

THIS FORMULA ONLY WORKS ON 2×2 MATRICES.
IF YOU WANT TO FIND THE INVERSE OF A BIGGER MATRIX, I'M AFRAID YOU'RE GOING TO HAVE TO SETTLE FOR GAUSSIAN ELIMINATION.
HMM
THAT'S TOO BAD...

NEXT, I THOUGHT I'D SHOW YOU HOW TO DETERMINE WHETHER A MATRIX HAS AN INVERSE OR NOT.
SO...SOME MATRICES LACK AN INVERSE?

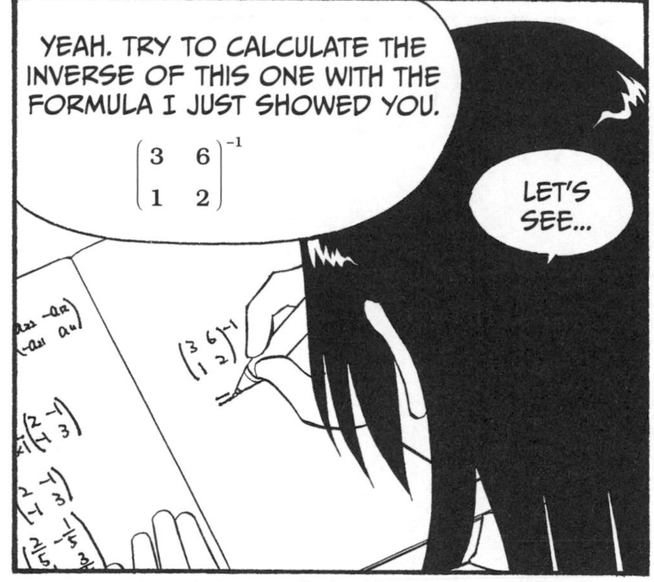
YEAH. TRY TO CALCULATE THE INVERSE OF THIS ONE WITH THE FORMULA I JUST SHOWED YOU.
$\begin{pmatrix} 3 & 6 \\ 1 & 2 \end{pmatrix}^{-1}$
LET'S SEE...

$\begin{pmatrix} 3 & 6 \\ 1 & 2 \end{pmatrix}^{-1} = \frac{1}{3 \cdot 2 - 6 \cdot 1} \begin{pmatrix} 2 & -6 \\ -1 & 3 \end{pmatrix}$
OH, THE DENOMINATOR BECOMES ZERO. I GUESS YOU'RE RIGHT.

ONE LAST THING: THE INVERSE OF AN INVERTIBLE MATRIX IS, OF COURSE, ALSO INVERTIBLE.
INVERTIBLE
$\begin{pmatrix} 3 & 1 \\ 1 & 2 \end{pmatrix} \begin{pmatrix} \frac{2}{5} & -\frac{1}{5} \\ -\frac{1}{5} & \frac{3}{5} \end{pmatrix} = \begin{pmatrix} 1 & 0 \\ 0 & 1 \end{pmatrix}$
NOT INVERTIBLE
$\begin{pmatrix} 3 & 6 \\ 1 & 2 \end{pmatrix}$
MAKES SENSE!

NOW FOR THE TEST TO SEE WHETHER A MATRIX IS INVERTIBLE OR NOT.

$$\det\begin{pmatrix} a_{11} & a_{12} & \cdots & a_{1n} \\ a_{21} & a_{22} & \cdots & a_{2n} \\ \vdots & \vdots & \ddots & \vdots \\ a_{n1} & a_{n2} & \cdots & a_{nn} \end{pmatrix}$$

IT'S ALSO WRITTEN WITH STRAIGHT BARS, LIKE THIS:

$$\begin{vmatrix} a_{11} & a_{12} & \cdots & a_{1n} \\ a_{21} & a_{22} & \cdots & a_{2n} \\ \vdots & \vdots & \ddots & \vdots \\ a_{n1} & a_{n2} & \cdots & a_{nn} \end{vmatrix}$$

DET?

determinant

IT'S SHORT FOR *DETERMINANT*.

DOES A GIVEN MATRIX HAVE AN INVERSE?

$$\det\begin{pmatrix} a_{11} & a_{12} & \cdots & a_{1n} \\ a_{21} & a_{22} & \cdots & a_{2n} \\ \vdots & \vdots & \ddots & \vdots \\ a_{n1} & a_{n2} & \cdots & a_{nn} \end{pmatrix} \neq 0 \quad \text{means that} \quad \begin{pmatrix} a_{11} & a_{12} & \cdots & a_{1n} \\ a_{21} & a_{22} & \cdots & a_{2n} \\ \vdots & \vdots & \ddots & \vdots \\ a_{n1} & a_{n2} & \cdots & a_{nn} \end{pmatrix}^{-1} \quad \text{exists.}$$

CALCULATING DETERMINANTS

$n=2$ $\begin{pmatrix} a_{11} & a_{12} \\ a_{21} & a_{22} \end{pmatrix}$

$n=3$ $\begin{pmatrix} a_{11} & a_{12} & a_{13} \\ a_{21} & a_{22} & a_{23} \\ a_{31} & a_{32} & a_{33} \end{pmatrix}$

THERE ARE SEVERAL DIFFERENT WAYS TO CALCULATE A DETERMINANT. WHICH ONE'S BEST DEPENDS ON THE SIZE OF THE MATRIX.

LET'S START WITH THE FORMULA FOR TWO-DIMENSIONAL MATRICES AND WORK OUR WAY UP.

SOUNDS GOOD.

TO FIND THE DETERMINANT OF A 2×2 MATRIX, JUST SUBSTITUTE THE EXPRESSION LIKE THIS.

$$\det\begin{pmatrix} a_{11} & a_{12} \\ a_{21} & a_{22} \end{pmatrix} = a_{11}a_{22} - a_{12}a_{21}$$

LET'S SEE WHETHER $\begin{pmatrix} 3 & 0 \\ 0 & 2 \end{pmatrix}$ HAS AN INVERSE OR NOT.

$$\det\begin{pmatrix} 3 & 0 \\ 0 & 2 \end{pmatrix} = 3 \cdot 2 - 0 \cdot 0 = 6$$

IT DOES, SINCE $\det\begin{pmatrix} 3 & 0 \\ 0 & 2 \end{pmatrix} \neq 0$.

INCIDENTALLY, THE AREA OF THE PARALLELOGRAM SPANNED BY THE FOLLOWING FOUR POINTS...

- THE ORIGIN
- THE POINT (a_{11}, a_{21})
- THE POINT (a_{12}, a_{22})
- THE POINT $(a_{11} + a_{12}, a_{21} + a_{22})$

...COINCIDES WITH THE ABSOLUTE VALUE OF

$$\text{dct}\begin{pmatrix} a_{11} & a_{12} \\ a_{21} & a_{22} \end{pmatrix}$$

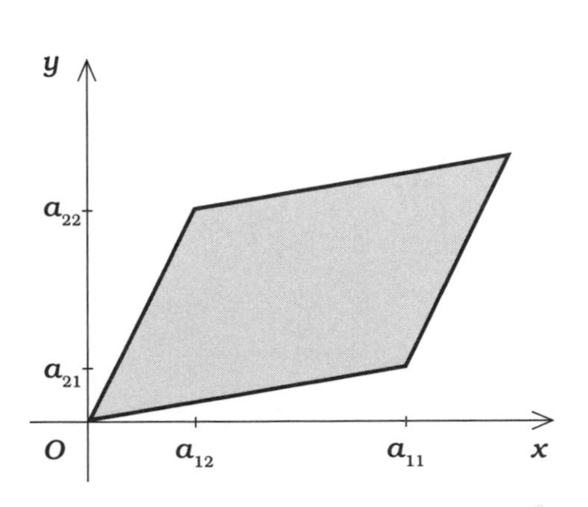

TO FIND THE DETERMINANT OF A 3×3 MATRIX, JUST USE THE FOLLOWING FORMULA.

THIS IS SOMETIMES CALLED *SARRUS' RULE.*

$$\det\begin{pmatrix} a_{11} & a_{12} & a_{13} \\ a_{21} & a_{22} & a_{23} \\ a_{31} & a_{32} & a_{33} \end{pmatrix} = a_{11}a_{22}a_{33} + a_{12}a_{23}a_{31} + a_{13}a_{21}a_{32} - a_{13}a_{22}a_{31} - a_{12}a_{21}a_{33} - a_{11}a_{23}a_{32}$$

SARRUS' RULE

Write out the matrix, and then write its first two columns again after the third column, giving you a total of five columns. Add the products of the diagonals going from top to bottom (indicated by the solid lines) and subtract the products of the diagonals going from bottom to top (indicated by dotted lines). This will generate the formula for Sarrus' Rule, and it's much easier to remember!

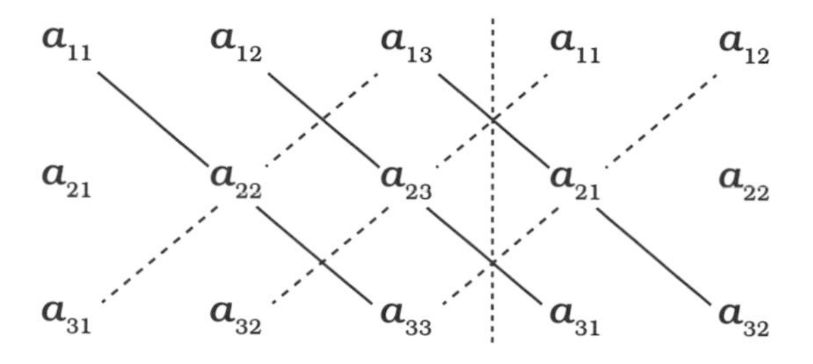

LET'S SEE IF $\begin{pmatrix} 1 & 0 & 0 \\ 1 & 1 & -1 \\ -2 & 0 & 3 \end{pmatrix}$ HAS AN INVERSE.

$$\det\begin{pmatrix} 1 & 0 & 0 \\ 1 & 1 & -1 \\ -2 & 0 & 3 \end{pmatrix} = 1 \cdot 1 \cdot 3 + 0 \cdot (-1) \cdot (-2) + 0 \cdot 1 \cdot 0 - 0 \cdot 1 \cdot (-2) - 0 \cdot 1 \cdot 3 - 1 \cdot (-1) \cdot 0$$

$$= 3 + 0 + 0 - 0 - 0 - 0$$

$$= 3$$

$$\det\begin{pmatrix} 1 & 0 & 0 \\ 1 & 1 & -1 \\ -2 & 0 & 3 \end{pmatrix} \neq 0$$

SO THIS ONE HAS AN INVERSE TOO!

AND THE VOLUME OF THE PARALLELEPIPED* SPANNED BY THE FOLLOWING EIGHT POINTS...

- THE ORIGIN
- THE POINT (a_{11}, a_{21}, a_{31})
- THE POINT (a_{12}, a_{22}, a_{32})
- THE POINT (a_{13}, a_{23}, a_{33})
- THE POINT $(a_{11} + a_{12}, a_{21} + a_{22}, a_{31} + a_{32})$
- THE POINT $(a_{11} + a_{13}, a_{21} + a_{23}, a_{31} + a_{33})$
- THE POINT $(a_{12} + a_{13}, a_{22} + a_{23}, a_{32} + a_{33})$
- THE POINT $(a_{11} + a_{12} + a_{13}, a_{21} + a_{22} + a_{23}, a_{31} + a_{32} + a_{33})$

...ALSO COINCIDES WITH THE ABSOLUTE VALUE OF

$$\det\begin{pmatrix} a_{11} & a_{12} & a_{13} \\ a_{21} & a_{22} & a_{23} \\ a_{31} & a_{32} & a_{33} \end{pmatrix}$$

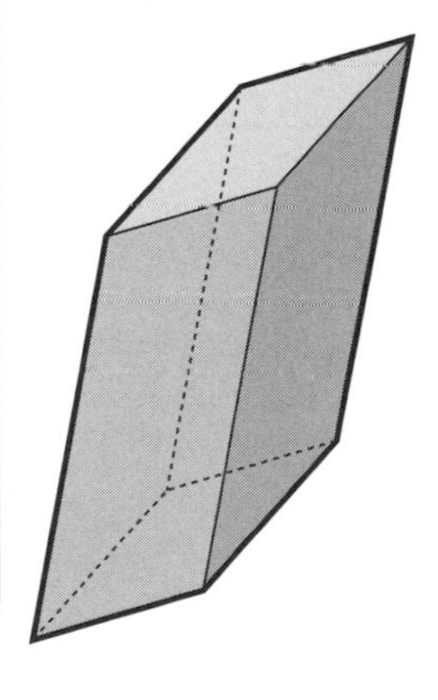

EACH PAIR OF OPPOSITE FACES ON THE PARALLELEPIPED ARE PARALLEL AND HAVE THE SAME AREA.

* A *PARALLELEPIPED* IS A THREE-DIMENSIONAL FIGURE FORMED BY SIX PARALLELOGRAMS.

SO NEXT UP ARE 4×4 MATRICES, I SUPPOSE...
a11 a12 a13 a14
a21 a22 a23 a24
a31 a32 a33 a34
a41 a42 a43 a44
YEP.

TEE HEE
MORE OF THIS PERHAPS?

I'M AFRAID NOT... THE GRIM TRUTH IS THAT THE FORMULAS USED TO CALCULATE DETERMINANTS OF DIMENSIONS FOUR AND ABOVE ARE VERY COMPLICATED.
NOPE!

SO HOW DO WE CALCULATE THEM?
?
TO BE ABLE TO DO THAT...

YEP, THE TERMS IN THE DETERMINANT FORMULA ARE FORMED ACCORDING TO CERTAIN RULES.

FLLLIP

TAKE A CLOSER LOOK AT THE TERM INDEXES.

?

$$\det\begin{pmatrix} a_{11} & a_{12} \\ a_{21} & a_{22} \end{pmatrix} = a_{11}a_{22} - a_{12}a_{21}$$

$$\det\begin{pmatrix} a_{11} & a_{12} & a_{13} \\ a_{21} & a_{22} & a_{23} \\ a_{31} & a_{32} & a_{33} \end{pmatrix} = a_{11}a_{22}a_{33} + a_{12}a_{23}a_{31} + a_{13}a_{21}a_{32} - a_{13}a_{22}a_{31} - a_{12}a_{21}a_{33} - a_{11}a_{23}a_{32}$$

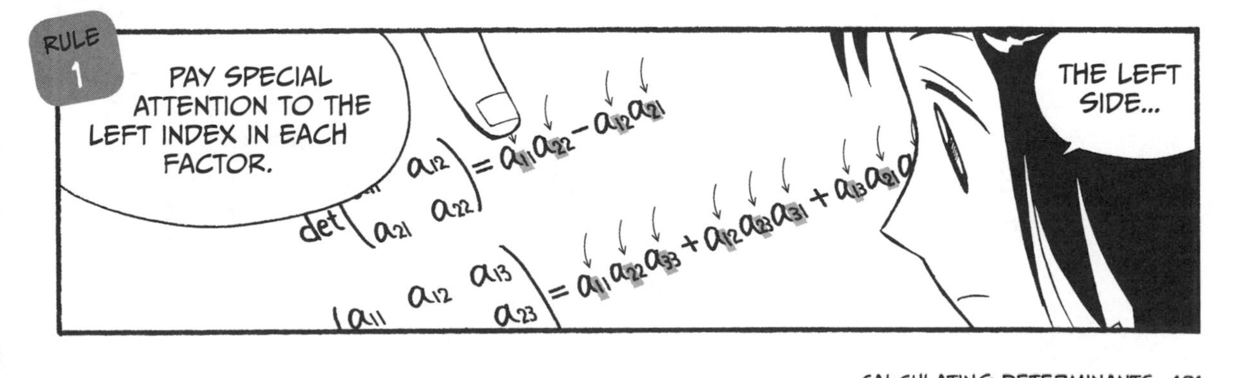

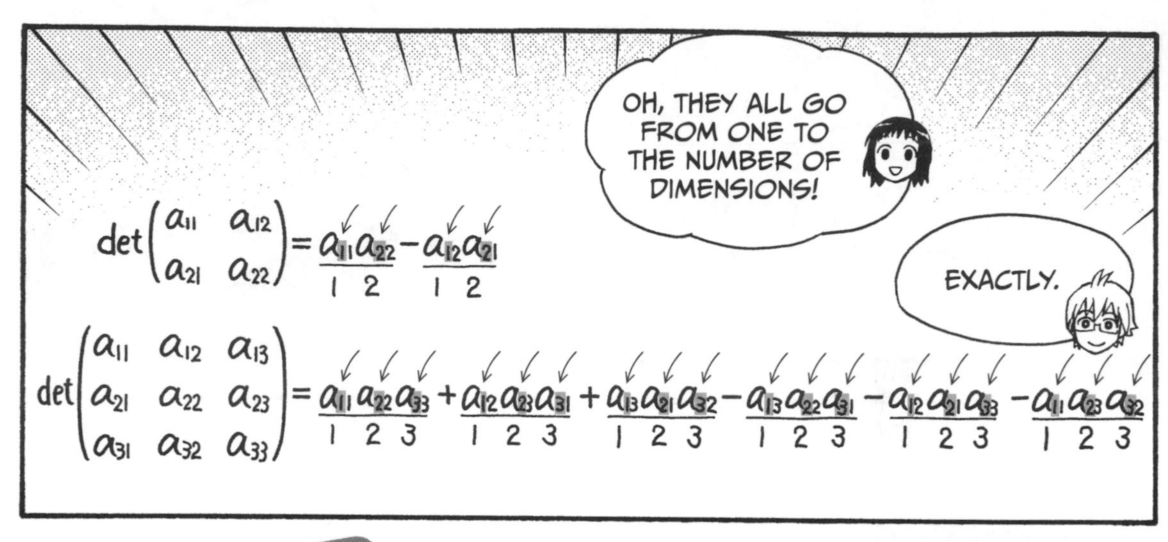

$$\det\begin{pmatrix} a_{11} & a_{12} \\ a_{21} & a_{22} \end{pmatrix} = \underset{1\ 2}{a_{11}a_{22}} - \underset{1\ 2}{a_{12}a_{21}}$$

$$\det\begin{pmatrix} a_{11} & a_{12} & a_{13} \\ a_{21} & a_{22} & a_{23} \\ a_{31} & a_{32} & a_{33} \end{pmatrix} = \underset{1\ 2\ 3}{a_{11}a_{22}a_{33}} + \underset{1\ 2\ 3}{a_{12}a_{23}a_{31}} + \underset{1\ 2\ 3}{a_{13}a_{21}a_{32}} - \underset{1\ 2\ 3}{a_{13}a_{22}a_{31}} - \underset{1\ 2\ 3}{a_{12}a_{21}a_{33}} - \underset{1\ 2\ 3}{a_{11}a_{23}a_{32}}$$

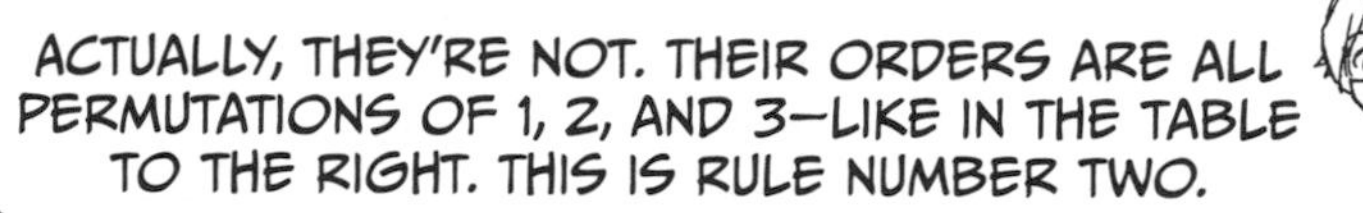

$$\underset{1\ 2}{a_{11}a_{22}} - \underset{2\ 1}{a_{12}a_{21}}$$

$$\underset{1\ 2\ 3}{a_{11}a_{22}a_{33}} + \underset{2\ 3\ 1}{a_{12}a_{23}a_{31}} + \underset{3\ 1\ 2}{a_{13}a_{21}a_{32}} - \underset{3\ 2\ 1}{a_{13}a_{22}a_{31}} - \underset{2\ 1\ 3}{a_{12}a_{21}a_{33}} - \underset{1\ 3\ 2}{a_{11}a_{23}a_{32}}$$

PERMUTATIONS OF 1–2

PATTERN 1	1 2
PATTERN 2	2 1

PERMUTATIONS OF 1–3

PATTERN 1	1 2 3
PATTERN 2	1 3 2
PATTERN 3	2 1 3
PATTERN 4	2 3 1
PATTERN 5	3 1 2
PATTERN 6	3 2 1

THE NEXT STEP IS TO FIND ALL THE PLACES WHERE TWO TERMS AREN'T IN THE NATURAL ORDER—MEANING THE PLACES WHERE TWO INDEXES HAVE TO BE SWITCHED FOR THEM TO BE IN AN INCREASING ORDER.

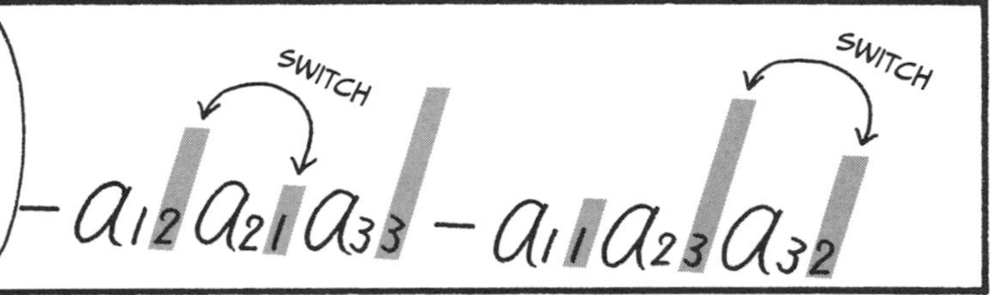

WE GATHER ALL THIS INFORMATION INTO A TABLE LIKE THIS.

	PERMUTATIONS OF 1–2	CORRESPONDING TERM IN THE DETERMINANT	SWITCHES
PATTERN 1	1 2	$a_{11} a_{22}$	
PATTERN 2	2 1	$a_{12} a_{21}$	2 AND 1

	PERMUTATIONS OF 1–3	CORRESPONDING TERM IN THE DETERMINANT	SWITCHES		
PATTERN 1	1 2 3	$a_{11} a_{22} a_{33}$			
PATTERN 2	1 3 2	$a_{11} a_{23} a_{32}$			3 AND 2
PATTERN 3	2 1 3	$a_{12} a_{21} a_{33}$	2 AND 1		
PATTERN 4	2 3 1	$a_{12} a_{23} a_{31}$	2 AND 1	3 AND 1	
PATTERN 5	3 1 2	$a_{13} a_{21} a_{32}$		3 AND 1	3 AND 2
PATTERN 6	3 2 1	$a_{13} a_{22} a_{31}$	2 AND 1	3 AND 1	3 AND 2

THEN WE COUNT HOW MANY SWITCHES WE NEED FOR EACH TERM.

IF THE NUMBER IS EVEN, WE WRITE THE TERM AS POSITIVE. IF IT IS ODD, WE WRITE IT AS NEGATIVE.

	PERMUTATIONS OF 1–2	CORRESPONDING TERM IN THE DETERMINANT	SWITCHES	NUMBER OF SWITCHES	SIGN
PATTERN 1	1 2	$a_{11}a_{22}$		0	+
PATTERN 2	2 1	$a_{12}a_{21}$	2 AND 1	1	−

	PERMUTATIONS OF 1–3	CORRESPONDING TERM IN THE DETERMINANT	SWITCHES			NUMBER OF SWITCHES	SIGN
PATTERN 1	1 2 3	$a_{11}a_{22}a_{33}$				0	+
PATTERN 2	1 3 2	$a_{11}a_{23}a_{32}$			3 AND 2	1	−
PATTERN 3	2 1 3	$a_{12}a_{21}a_{33}$	2 AND 1			1	−
PATTERN 4	2 3 1	$a_{12}a_{23}a_{31}$	2 AND 1	3 AND 1		2	+
PATTERN 5	3 1 2	$a_{13}a_{21}a_{32}$		3 AND 1	3 AND 2	2	+
PATTERN 6	3 2 1	$a_{13}a_{22}a_{31}$	2 AND 1	3 AND 1	3 AND 2	3	−

LIKE THIS.

HMM...

TRY COMPARING OUR EARLIER DETERMINANT FORMULAS WITH THE COLUMNS "CORRESPONDING TERM IN THE DETERMINANT" AND "SIGN."

$$\det\begin{pmatrix} a_{11} & a_{12} \\ a_{21} & a_{22} \end{pmatrix} = a_{11}a_{22} - a_{12}a_{21}$$

CORRESPONDING TERM IN THE DETERMINANT	SIGN
$a_{11}a_{22}$	+
$a_{12}a_{21}$	−

$$\det\begin{pmatrix} a_{11} & a_{12} & a_{13} \\ a_{21} & a_{22} & a_{23} \\ a_{31} & a_{32} & a_{33} \end{pmatrix} = a_{11}a_{22}a_{33} + a_{12}a_{23}a_{31} + a_{13}a_{21}a_{32} - a_{13}a_{22}a_{31} - a_{12}a_{21}a_{33} - a_{11}a_{23}a_{32}$$

CORRESPONDING TERM IN THE DETERMINANT	SIGN
$a_{11}a_{22}a_{33}$	+
$a_{11}a_{23}a_{32}$	−
$a_{12}a_{21}a_{33}$	−
$a_{12}a_{23}a_{31}$	+
$a_{13}a_{21}a_{32}$	+
$a_{13}a_{22}a_{31}$	−

WOW, THEY'RE THE SAME!

EXACTLY, AND THAT'S THE THIRD RULE.

	PERMUTATIONS OF 1–4	CORRESPONDING TERM IN THE DETERMINANT	SWITCHES						NUM. OF SWITCHES	SIGN
PATTERN 1	1 2 3 4	a_{11} a_{22} a_{33} a_{44}							0	+
PATTERN 2	1 2 4 3	a_{11} a_{22} a_{34} a_{43}						4 & 3	1	−
PATTERN 3	1 3 2 4	a_{11} a_{23} a_{32} a_{44}			3 & 2				1	−
PATTERN 4	1 3 4 2	a_{11} a_{23} a_{34} a_{42}			3 & 2		4 & 2		2	+
PATTERN 5	1 4 2 3	a_{11} a_{24} a_{32} a_{43}					4 & 2	4 & 3	2	+
PATTERN 6	1 4 3 2	a_{11} a_{24} a_{33} a_{42}			3 & 2		4 & 2	4 & 3	3	−
PATTERN 7	2 1 3 4	a_{12} a_{21} a_{33} a_{44}	2 & 1						1	−
PATTERN 8	2 1 4 3	a_{12} a_{21} a_{34} a_{43}	2 & 1					4 & 3	2	+
PATTERN 9	2 3 1 4	a_{12} a_{23} a_{31} a_{44}	2 & 1	3 & 1					2	+
PATTERN 10	2 3 4 1	a_{12} a_{23} a_{34} a_{41}	2 & 1	3 & 1		4 & 1			3	−
PATTERN 11	2 4 1 3	a_{12} a_{24} a_{31} a_{43}	2 & 1			4 & 1		4 & 3	3	−
PATTERN 12	2 4 3 1	a_{12} a_{24} a_{33} a_{41}	2 & 1	3 & 1		4 & 1		4 & 3	4	+
PATTERN 13	3 1 2 4	a_{13} a_{21} a_{32} a_{44}		3 & 1	3 & 2				2	+
PATTERN 14	3 1 4 2	a_{13} a_{21} a_{34} a_{42}		3 & 1	3 & 2		4 & 2		3	−
PATTERN 15	3 2 1 4	a_{13} a_{22} a_{31} a_{44}	2 & 1	3 & 1	3 & 2				3	−
PATTERN 16	3 2 4 1	a_{13} a_{22} a_{34} a_{41}	2 & 1	3 & 1	3 & 2	4 & 1			4	+
PATTERN 17	3 4 1 2	a_{13} a_{24} a_{31} a_{42}		3 & 1	3 & 2	4 & 1	4 & 2		4	+
PATTERN 18	3 4 2 1	a_{13} a_{24} a_{32} a_{41}	2 & 1	3 & 1	3 & 2	4 & 1	4 & 2		5	−
PATTERN 19	4 1 2 3	a_{14} a_{21} a_{32} a_{43}				4 & 1	4 & 2	4 & 3	3	−
PATTERN 20	4 1 3 2	a_{14} a_{21} a_{33} a_{42}			3 & 2	4 & 1	4 & 2	4 & 3	4	+
PATTERN 21	4 2 1 3	a_{14} a_{22} a_{31} a_{43}	2 & 1			4 & 1	4 & 2	4 & 3	4	+
PATTERN 22	4 2 3 1	a_{14} a_{22} a_{33} a_{41}	2 & 1	3 & 1		4 & 1	4 & 2	4 & 3	5	−
PATTERN 23	4 3 1 2	a_{14} a_{23} a_{31} a_{42}		3 & 1	3 & 2	4 & 1	4 & 2	4 & 3	5	−
PATTERN 24	4 3 2 1	a_{14} a_{23} a_{32} a_{41}	2 & 1	3 & 1	3 & 2	4 & 1	4 & 2	4 & 3	6	+

IF THIS IS ON THE TEST, I'M DONE FOR...

DON'T WORRY, MOST TEACHERS WILL GIVE YOU PROBLEMS INVOLVING ONLY 2×2 AND 3×3 MATRICES.
PHEW
I HOPE SO...

I THINK THAT'S ENOUGH FOR TODAY. WE GOT THROUGH A TON OF NEW MATERIAL.
THANKS, REIJI. YOU'RE THE BEST!

TIME REALLY FLEW BY, THOUGH...
...

MAYBE I CAN...

お会計 CASHE
DO YOU NEED A BAG FOR THAT?
JEEZ, LOOK AT THE TIME...
POW!
THERE SHOULDN'T BE ANYONE LEFT IN THERE AT THIS HOUR.
ガン
SMACK!
ガン
BIFF!

真派花道

CALCULATING INVERSE MATRICES USING COFACTORS

There are two practical ways to calculate inverse matrices, as mentioned on page 88.

- Using cofactors
- Using Gaussian elimination

Since the cofactor method involves a lot of cumbersome calculations, we avoided using it in this chapter. However, since most books seem to introduce the method, here's a quick explanation.

To use this method, you first have to understand these two concepts:

- The (i, j)-minor, written as M_{ij}
- The (i, j)-cofactor, written as C_{ij}

So first we'll have a look at these.

M_{ij}

The (i, j)-minor is the determinant produced when we remove row i and column j from the $n \times n$ matrix A:

$$M_{ij} = \det \begin{pmatrix} a_{11} & a_{12} & \cdots & a_{1j} & \cdots & a_{1n} \\ a_{21} & a_{22} & \cdots & a_{2j} & \cdots & a_{2n} \\ \vdots & \vdots & \ddots & \vdots & & \vdots \\ a_{i1} & a_{i2} & \cdots & a_{ij} & \cdots & a_{in} \\ \vdots & \vdots & & \vdots & \ddots & \vdots \\ a_{n1} & a_{n2} & \cdots & a_{nj} & \cdots & a_{nn} \end{pmatrix}$$

All the minors of the 3×3 matrix $\begin{pmatrix} 1 & 0 & 0 \\ 1 & 1 & -1 \\ -2 & 0 & 3 \end{pmatrix}$ are listed on the next page.

M_{11} (1, 1)	M_{12} (1, 2)	M_{13} (1, 3)
$\det\begin{pmatrix} 1 & -1 \\ 0 & 3 \end{pmatrix} = 3$	$\det\begin{pmatrix} 1 & -1 \\ -2 & 3 \end{pmatrix} = 1$	$\det\begin{pmatrix} 1 & 1 \\ -2 & 0 \end{pmatrix} = 2$
M_{21} (2, 1)	M_{22} (2, 2)	M_{23} (2, 3)
$\det\begin{pmatrix} 0 & 0 \\ 0 & 3 \end{pmatrix} = 0$	$\det\begin{pmatrix} 1 & 0 \\ -2 & 3 \end{pmatrix} = 3$	$\det\begin{pmatrix} 1 & 0 \\ -2 & 0 \end{pmatrix} = 0$
M_{31} (3, 1)	M_{32} (3, 2)	M_{33} (3, 3)
$\det\begin{pmatrix} 0 & 0 \\ 1 & -1 \end{pmatrix} = 0$	$\det\begin{pmatrix} 1 & 0 \\ 1 & -1 \end{pmatrix} = -1$	$\det\begin{pmatrix} 1 & 0 \\ 1 & 1 \end{pmatrix} = 1$

C_{ij}

If we multiply the (i, j)-minor by $(-1)^{i+j}$, we get the (i, j)-cofactor. The standard way to write this is C_{ij}. The table below contains all cofactors of the 3×3 matrix

$$\begin{pmatrix} 1 & 0 & 0 \\ 1 & 1 & -1 \\ -2 & 0 & 3 \end{pmatrix}$$

C_{11} (1, 1)	C_{12} (1, 2)	C_{13} (1, 3)
$= (-1)^{1+1} \cdot \det\begin{pmatrix} 1 & -1 \\ 0 & 3 \end{pmatrix}$ $= 1 \cdot 3$ $= 3$	$= (-1)^{1+2} \cdot \det\begin{pmatrix} 1 & -1 \\ -2 & 3 \end{pmatrix}$ $= (-1) \cdot 1$ $= -1$	$= (-1)^{1+3} \cdot \det\begin{pmatrix} 1 & 1 \\ -2 & 0 \end{pmatrix}$ $= 1 \cdot 2$ $= 2$
C_{21} (2, 1)	C_{22} (2, 2)	C_{23} (2, 3)
$= (-1)^{2+1} \cdot \det\begin{pmatrix} 0 & 0 \\ 0 & 3 \end{pmatrix}$ $= (-1) \cdot 0$ $= 0$	$= (-1)^{2+2} \cdot \det\begin{pmatrix} 1 & 0 \\ -2 & 3 \end{pmatrix}$ $= 1 \cdot 3$ $= 3$	$= (-1)^{2+3} \cdot \det\begin{pmatrix} 1 & 0 \\ -2 & 0 \end{pmatrix}$ $= (-1) \cdot 0$ $= 0$
C_{31} (3, 1)	C_{32} (3, 2)	C_{33} (3, 3)
$= (-1)^{3+1} \cdot \det\begin{pmatrix} 0 & 0 \\ 1 & -1 \end{pmatrix}$ $= 1 \cdot 0$ $= 0$	$= (-1)^{3+2} \cdot \det\begin{pmatrix} 1 & 0 \\ 1 & -1 \end{pmatrix}$ $= (-1) \cdot (-1)$ $= 1$	$= (-1)^{3+3} \cdot \det\begin{pmatrix} 1 & 0 \\ 1 & 1 \end{pmatrix}$ $= 1 \cdot 1$ $= 1$

The $n \times n$ matrix

$$\begin{pmatrix} C_{11} & C_{21} & \cdots & C_{n1} \\ C_{12} & C_{22} & \cdots & C_{n2} \\ \vdots & \vdots & \ddots & \vdots \\ C_{1n} & C_{2n} & \cdots & C_{nn} \end{pmatrix}$$

which at place (i, j) has the (j, i)-cofactor[1] of the original matrix is called a *cofactor matrix*.

The sum of any row or column of the $n \times n$ matrix

$$\begin{pmatrix} a_{11}C_{11} & a_{21}C_{21} & \cdots & a_{n1}C_{n1} \\ a_{12}C_{12} & a_{22}C_{22} & \cdots & a_{n2}C_{n2} \\ \vdots & \vdots & \ddots & \vdots \\ a_{1n}C_{1n} & a_{2n}C_{2n} & \cdots & a_{nn}C_{nn} \end{pmatrix}$$

is equal to the determinant of the original $n \times n$ matrix

$$\begin{pmatrix} a_{11} & a_{12} & \cdots & a_{1n} \\ a_{21} & a_{22} & \cdots & a_{2n} \\ \vdots & \vdots & \ddots & \vdots \\ a_{n1} & a_{n2} & \cdots & a_{nn} \end{pmatrix}$$

CALCULATING INVERSE MATRICES

The inverse of a matrix can be calculated using the following formula:

$$\begin{pmatrix} a_{11} & a_{12} & \cdots & a_{1n} \\ a_{21} & a_{22} & \cdots & a_{2n} \\ \vdots & \vdots & \ddots & \vdots \\ a_{n1} & a_{n2} & \cdots & a_{nn} \end{pmatrix}^{-1} = \frac{1}{\det\begin{pmatrix} a_{11} & a_{12} & \cdots & a_{1n} \\ a_{21} & a_{22} & \cdots & a_{2n} \\ \vdots & \vdots & \ddots & \vdots \\ a_{n1} & a_{n2} & \cdots & a_{nn} \end{pmatrix}} \begin{pmatrix} C_{11} & C_{21} & \cdots & C_{n1} \\ C_{12} & C_{22} & \cdots & C_{n2} \\ \vdots & \vdots & \ddots & \vdots \\ C_{1n} & C_{2n} & \cdots & C_{nn} \end{pmatrix}$$

1. This is not a typo. (j, i)-cofactor is the correct index order. This is the transpose of the matrix with the cofactors in the expected positions.

For example, the inverse of the 3×3 matrix

$$\begin{pmatrix} 1 & 0 & 0 \\ 1 & 1 & -1 \\ -2 & 0 & 3 \end{pmatrix}$$

is equal to

$$\begin{pmatrix} 1 & 0 & 0 \\ 1 & 1 & -1 \\ -2 & 0 & 3 \end{pmatrix}^{-1} = \frac{1}{\det\begin{pmatrix} 1 & 0 & 0 \\ 1 & 1 & -1 \\ -2 & 0 & 3 \end{pmatrix}} \begin{pmatrix} 3 & 0 & 0 \\ -1 & 3 & 1 \\ 2 & 0 & 1 \end{pmatrix} = \frac{1}{3} \begin{pmatrix} 3 & 0 & 0 \\ -1 & 3 & 1 \\ 2 & 0 & 1 \end{pmatrix}$$

USING DETERMINANTS

The method presented in this chapter only defines the determinant and does nothing to explain what it is used for. A typical application (in image processing, for example) can easily reach determinant sizes in the $n = 100$ range, which with the approach used here would produce insurmountable numbers of calculations.

Because of this, determinants are usually calculated by first simplifying them with Gaussian elimination–like methods and then using these three properties, which can be derived using the definition presented in the book:

- If a row (or column) in a determinant is replaced by the sum of the row (column) and a multiple of another row (column), the value stays unchanged.
- If two rows (or columns) switch places, the values of the determinant are multiplied by –1.
- The value of an upper or lower triangular determinant is equal to the product of its main diagonal.

The difference between the two methods is so extreme that determinants that would be practically impossible to calculate (even using modern computers) with the first method can be done in a jiffy with the second one.

SOLVING LINEAR SYSTEMS WITH CRAMER'S RULE

Gaussian elimination, as presented on page 89, is only one of many methods you can use to solve linear systems. Even though Gaussian elimination is one of the best ways to solve them by hand, it is always good to know about alternatives, which is why we'll cover the *Cramer's rule* method next.

PROBLEM

Use Cramer's rule to solve the following linear system:

$$\begin{cases} 3x_1 + 1x_2 = 1 \\ 1x_1 + 2x_2 = 0 \end{cases}$$

SOLUTION

STEP 1 Rewrite the system

$$\begin{cases} a_{11}x_1 + a_{12}x_2 + \cdots + a_{1n}x_n = b_1 \\ a_{21}x_1 + a_{22}x_2 + \cdots + a_{2n}x_n = b_2 \\ \cdots\cdots\cdots\cdots\cdots\cdots\cdots\cdots\cdots \\ a_{n1}x_1 + a_{n2}x_2 + \cdots + a_{nn}x_n = b_n \end{cases}$$

like so:

$$\begin{pmatrix} a_{11} & a_{12} & \cdots & a_{1n} \\ a_{21} & a_{22} & \cdots & a_{2n} \\ \vdots & \vdots & \ddots & \vdots \\ a_{n1} & a_{n2} & \cdots & a_{nn} \end{pmatrix}\begin{pmatrix} x_1 \\ x_2 \\ \vdots \\ x_n \end{pmatrix} = \begin{pmatrix} b_1 \\ b_2 \\ \vdots \\ b_n \end{pmatrix}$$

If we rewrite

$$\begin{cases} 3x_1 + 1x_2 = 1 \\ 1x_1 + 2x_2 = 0 \end{cases}$$

we get

$$\begin{pmatrix} 3 & 1 \\ 1 & 2 \end{pmatrix}\begin{pmatrix} x_1 \\ x_2 \end{pmatrix} = \begin{pmatrix} 1 \\ 0 \end{pmatrix}$$

STEP 2 Make sure that

$$\det\begin{pmatrix} a_{11} & a_{12} & \cdots & a_{1n} \\ a_{21} & a_{22} & \cdots & a_{2n} \\ \vdots & \vdots & \ddots & \vdots \\ a_{n1} & a_{n2} & \cdots & a_{nn} \end{pmatrix} \neq 0$$

We have

$$\det\begin{pmatrix} 3 & 1 \\ 1 & 2 \end{pmatrix} = 3 \cdot 2 - 1 \cdot 1 \neq 0$$

STEP 3 Replace each column with the solution vector to get the corresponding solution:

Column i

$$x_i = \frac{\det\begin{pmatrix} a_{11} & a_{12} & \cdots & b_1 & \cdots & a_{1n} \\ a_{21} & a_{22} & \cdots & b_2 & \cdots & a_{2n} \\ \vdots & \vdots & \ddots & \vdots & \ddots & \vdots \\ a_{n1} & a_{n2} & \cdots & b_n & \cdots & a_{nn} \end{pmatrix}}{\det\begin{pmatrix} a_{11} & a_{12} & \cdots & a_{1i} & \cdots & a_{1n} \\ a_{21} & a_{22} & \cdots & a_{2i} & \cdots & a_{2n} \\ \vdots & \vdots & \ddots & \vdots & \ddots & \vdots \\ a_{n1} & a_{n2} & \cdots & a_{ni} & \cdots & a_{nn} \end{pmatrix}}$$

- $$x_1 = \frac{\det\begin{pmatrix} 1 & 1 \\ 0 & 2 \end{pmatrix}}{\det\begin{pmatrix} 3 & 1 \\ 1 & 2 \end{pmatrix}} = \frac{1 \cdot 2 - 1 \cdot 0}{5} = \frac{2}{5}$$
- $$x_2 = \frac{\det\begin{pmatrix} 3 & 1 \\ 1 & 0 \end{pmatrix}}{\det\begin{pmatrix} 3 & 1 \\ 1 & 2 \end{pmatrix}} = \frac{3 \cdot 0 - 1 \cdot 1}{5} = -\frac{1}{5}$$

5
INTRODUCTION TO VECTORS

ONE MORE MINUTE!
BAM
PANT
PANT
OSSU!
YOU CAN DO BETTER THAN THAT!
SMACK
PUT YOUR BACK INTO IT!
OSSU!
BAM
GREAT!
LET'S LEAVE IT AT THAT FOR TODAY.
WHEEZE
PANT
PANT
M-MORE!

I CAN DO MORE!

YOU CAN BARELY EVEN STAND!
THUMP
UWA!!

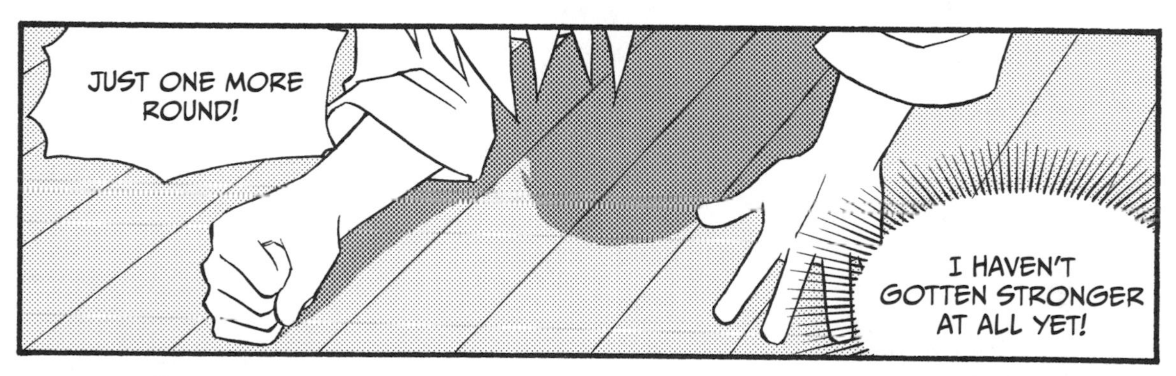
JUST ONE MORE ROUND!
I HAVEN'T GOTTEN STRONGER AT ALL YET!

HEHEH, FINE BY ME!
OSSU!

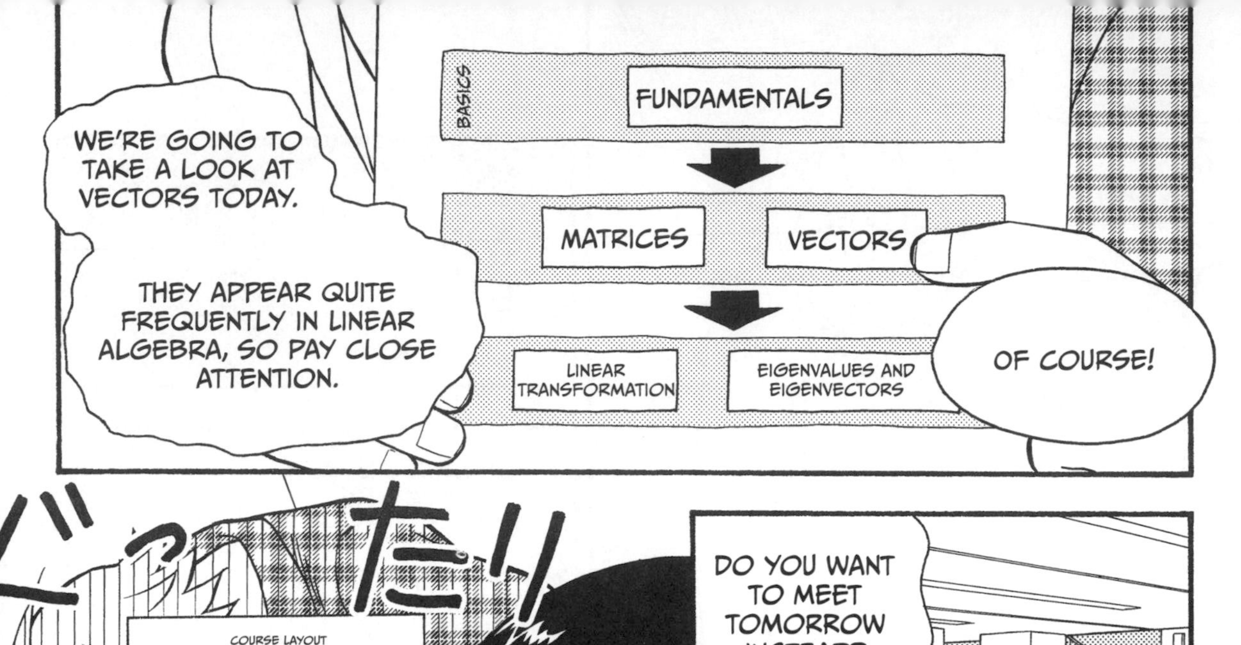

WHAT ARE VECTORS?

WHAT KIND OF INTER-PRETATION?
I THINK IT'LL BE EASIER TO EXPLAIN USING AN EXAMPLE...

MINIGOLF SHOULD MAKE AN EXCELLENT METAPHOR.
MINIGOLF?

DON'T BE INTIMIDATED—MY PUTTING SKILLS ARE PRETTY RUSTY.
GOAL
HAHA—MINE TOO!

STARTING POINT
x_2
4
O
7
x_1
OUR COURSE WILL LOOK LIKE THIS.
WE'LL USE COORDINATES TO DESCRIBE WHERE THE BALL AND HOLE ARE TO MAKE EXPLAINING EASIER.
THAT MEANS THAT THE STARTING POINT IS AT (0, 0) AND THAT THE HOLE IS AT (7, 4), RIGHT?

PLAYER 1
REIJI YURINO

I WENT FIRST. I PLAYED CONSERVATIVELY AND PUT THE BALL IN WITH THREE STROKES.

REPLAY

FIRST STROKE

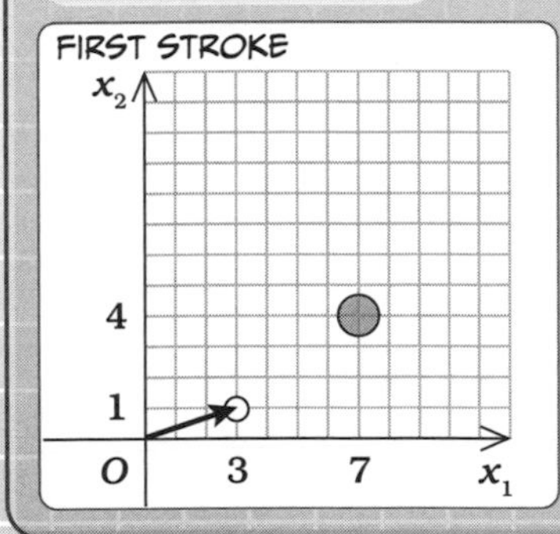

SECOND STROKE

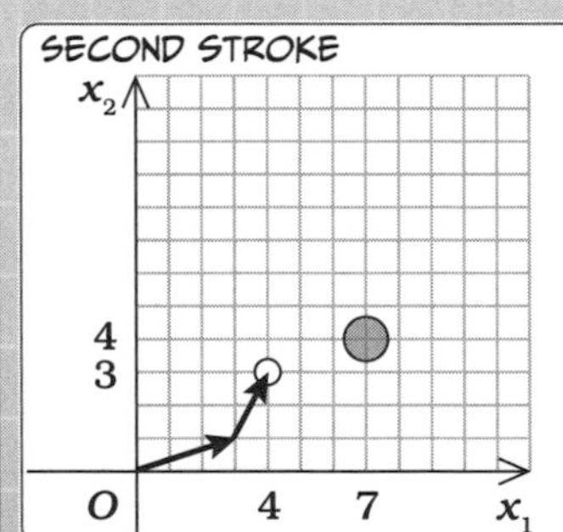

THIRD STROKE

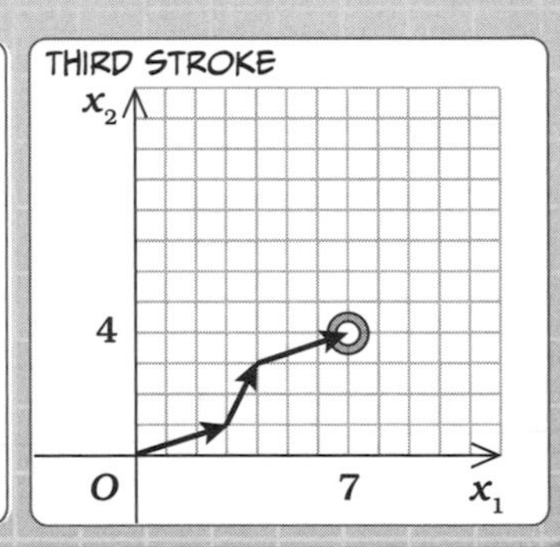

STROKE INFORMATION

	First stroke	Second stroke	Third stroke
Ball position	Point (3, 1)	Point (4, 3)	Point (7, 4)
Ball position relative to its last position	3 to the right and 1 up relative to (0, 0)	1 to the right and 2 up relative to (3, 1)	3 to the right and 1 up relative to (4, 3)
Ball movement expressed in the form (to the right, up)	(3, 1)	(3, 1) + (1, 2) = (4, 3)	(3, 1) + (1, 2) + (3, 1) = (7, 4)

PLAYER 2
MISA ICHINOSE

YOU GAVE THE BALL A GOOD WALLOP AND PUT THE BALL IN WITH TWO STROKES.

REPLAY

FIRST STROKE

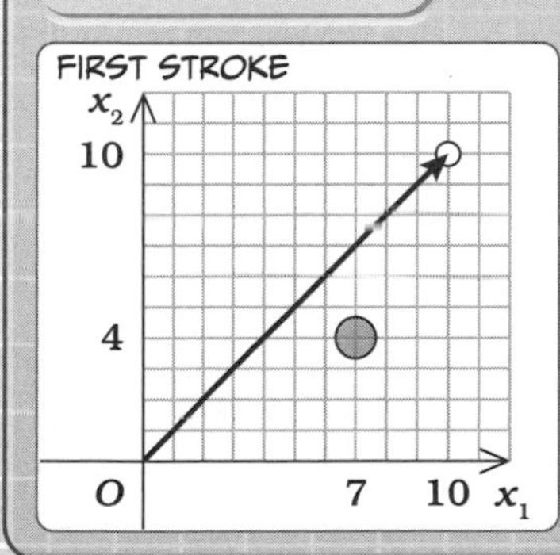

SECOND STROKE

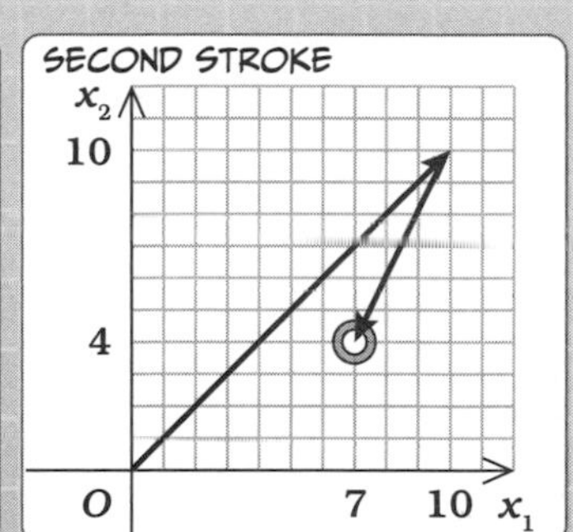

STROKE INFORMATION

	First stroke	Second stroke
Ball position	Point (10, 10)	Point (7, 4)
Ball position relative to its last position	10 to the right and 10 up relative to (0, 0)	–3 to the right and –6 up relative to (10, 10)
Ball movement expressed in the form (to the right, up)	(10, 10)	(10, 10) + (–3, –6) = (7, 4)

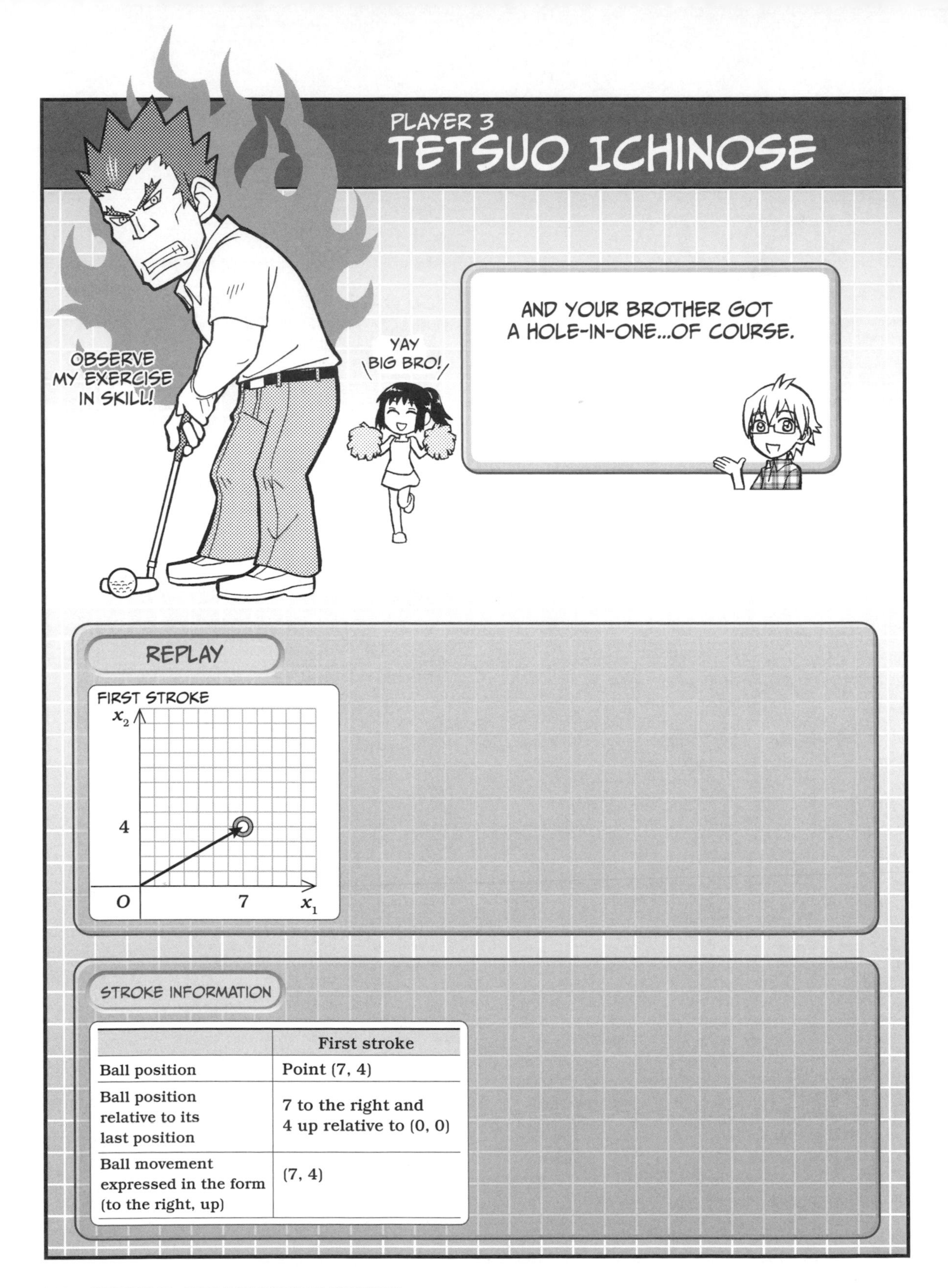

	First stroke
Ball position	Point (7, 4)
Ball position relative to its last position	7 to the right and 4 up relative to (0, 0)
Ball movement expressed in the form (to the right, up)	(7, 4)

WELL, AT LEAST WE ALL MADE IT IN!
START
GOAL
TRY TO REMEMBER THE MINIGOLF EXAMPLE WHILE WE TALK ABOUT THE NEXT FEW SUBJECTS.
1×n matrices $(a_1 \ a_2 \ \cdots \ a_n)$ and n×1 matrices $\begin{pmatrix} a_1 \\ a_2 \\ \vdots \\ a_n \end{pmatrix}$
VECTORS CAN BE INTERPRETED IN FOUR DIFFERENT WAYS. LET ME GIVE YOU A QUICK WALK-THROUGH OF ALL OF THEM.

I'LL USE THE 1×2 MATRIX (7, 4) AND THE 2×1 MATRIX $\begin{pmatrix} 7 \\ 4 \end{pmatrix}$ TO MAKE THINGS SIMPLER.
OKAY.

INTERPRETATION 1

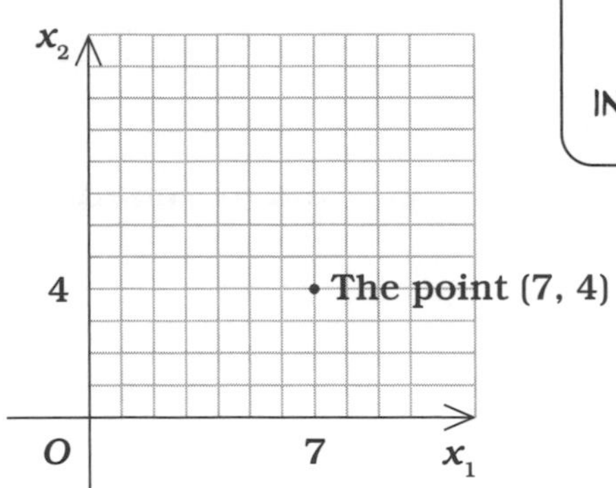

INTERPRETATION 2

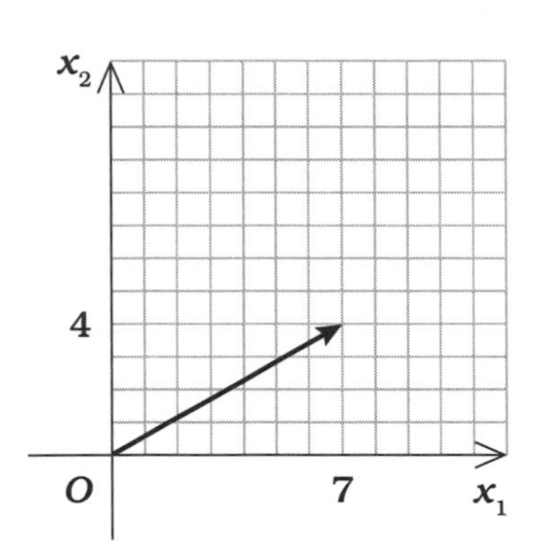

INTERPRETATION 3

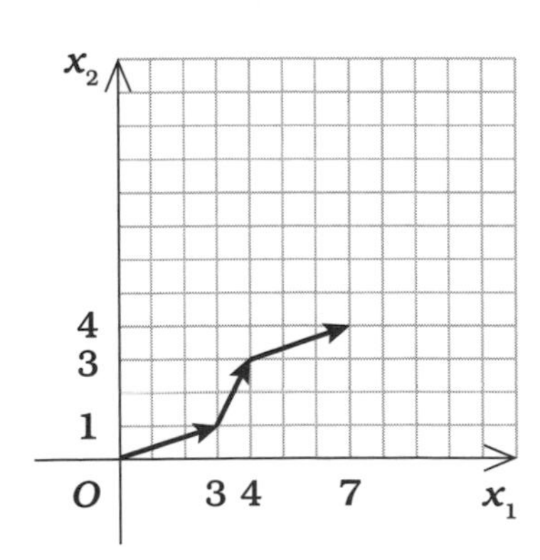

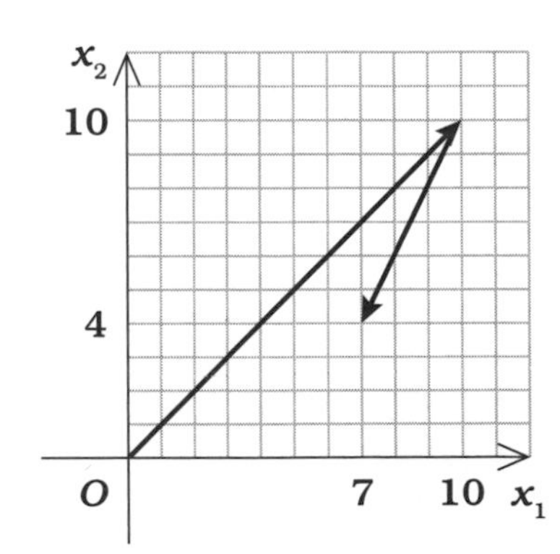

INTERPRETATION 4

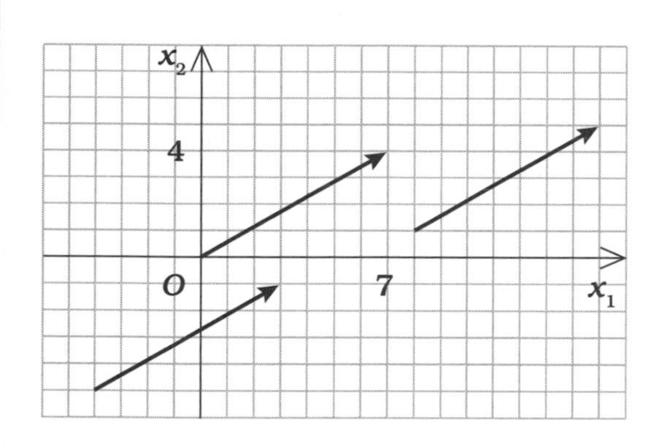

VECTORS ARE KIND OF A MYSTERIOUS CONCEPT, DON'T YOU THINK?
WELL, THEY MAY SEEM THAT WAY AT FIRST.
BUT...
ONCE YOU'VE GOT THE BASICS DOWN, YOU'LL BE ABLE TO APPLY THEM TO ALL SORTS OF INTERESTING PROBLEMS.
FOR EXAMPLE, THEY'RE FREQUENTLY USED IN PHYSICS TO DESCRIBE DIFFERENT TYPES OF FORCES.
COOL!
PUTTING FORCE
GRAVITATIONAL FORCE

VECTOR CALCULATIONS

EVEN THOUGH VECTORS HAVE A FEW SPECIAL INTERPRETATIONS, THEY'RE ALL JUST $1 \times n$ AND $n \times 1$ MATRICES...

ADDITION

- $(10, 10) + (-3, -6) = (10 + (-3), 10 + (-6)) = (7, 4)$
- $\begin{pmatrix} 10 \\ 10 \end{pmatrix} + \begin{pmatrix} -3 \\ -6 \end{pmatrix} = \begin{pmatrix} 10 + (-3) \\ 10 + (-6) \end{pmatrix} = \begin{pmatrix} 7 \\ 4 \end{pmatrix}$

SUBTRACTION

- $(10, 10) - (3, 6) = (10 - 3, 10 - 6) = (7, 4)$
- $\begin{pmatrix} 10 \\ 10 \end{pmatrix} - \begin{pmatrix} 3 \\ 6 \end{pmatrix} = \begin{pmatrix} 10 - 3 \\ 10 - 6 \end{pmatrix} = \begin{pmatrix} 7 \\ 4 \end{pmatrix}$

SCALAR MULTIPLICATION

- $2(3, 1) = (2 \cdot 3, 2 \cdot 1) = (6, 2)$
- $2\begin{pmatrix} 3 \\ 1 \end{pmatrix} = \begin{pmatrix} 2 \cdot 3 \\ 2 \cdot 1 \end{pmatrix} = \begin{pmatrix} 6 \\ 2 \end{pmatrix}$

MATRIX MULTIPLICATION

- $\begin{pmatrix} 3 \\ 1 \end{pmatrix}(1, 2) = \begin{pmatrix} 3 \cdot 1 & 3 \cdot 2 \\ 1 \cdot 1 & 1 \cdot 2 \end{pmatrix} = \begin{pmatrix} 3 & 6 \\ 1 & 2 \end{pmatrix}$
- $(3, 1)\begin{pmatrix} 1 \\ 2 \end{pmatrix} = (3 \cdot 1 + 1 \cdot 2) = 5$
- $\begin{pmatrix} 8 & -3 \\ 2 & 1 \end{pmatrix}\begin{pmatrix} 3 \\ 1 \end{pmatrix} = \begin{pmatrix} 8 \cdot 3 + (-3) \cdot 1 \\ 2 \cdot 3 + 1 \cdot 1 \end{pmatrix} = \begin{pmatrix} 21 \\ 7 \end{pmatrix} = 7\begin{pmatrix} 3 \\ 1 \end{pmatrix}$

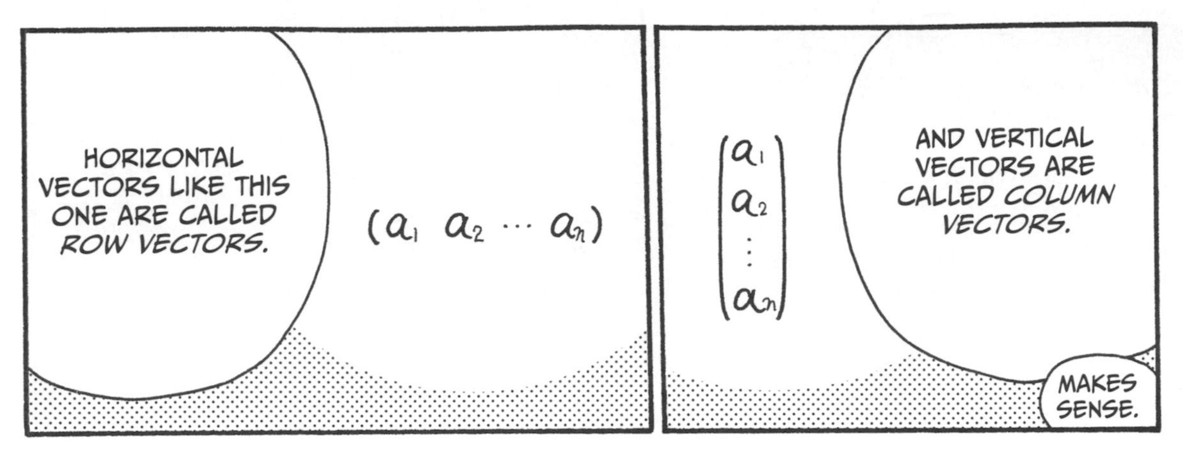
HORIZONTAL VECTORS LIKE THIS ONE ARE CALLED *ROW VECTORS.*
$(a_1 \; a_2 \; \cdots \; a_n)$
AND VERTICAL VECTORS ARE CALLED *COLUMN VECTORS.*
$\begin{pmatrix} a_1 \\ a_2 \\ \vdots \\ a_n \end{pmatrix}$
MAKES SENSE.

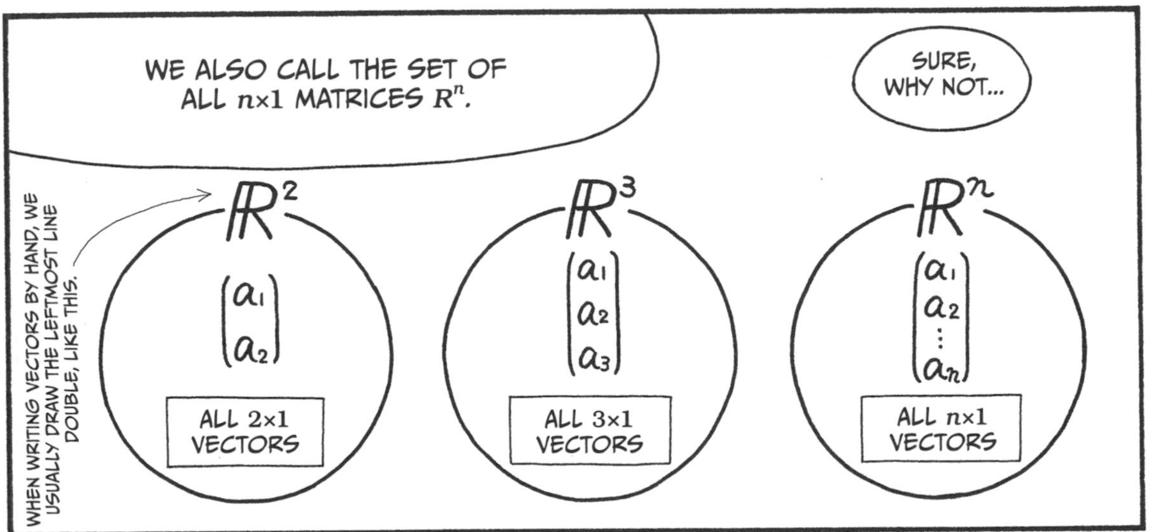
WE ALSO CALL THE SET OF ALL $n\times1$ MATRICES R^n.
SURE, WHY NOT...
WHEN WRITING VECTORS BY HAND, WE USUALLY DRAW THE LEFTMOST LINE DOUBLE, LIKE THIS.
$\mathbb{R}^2$
$\begin{pmatrix} a_1 \\ a_2 \end{pmatrix}$
ALL 2×1 VECTORS
$\mathbb{R}^3$
$\begin{pmatrix} a_1 \\ a_2 \\ a_3 \end{pmatrix}$
ALL 3×1 VECTORS
$\mathbb{R}^n$
$\begin{pmatrix} a_1 \\ a_2 \\ \vdots \\ a_n \end{pmatrix}$
ALL $n\times1$ VECTORS

R^n APPEARS A LOT IN LINEAR ALGEBRA, SO MAKE SURE YOU REMEMBER IT.
NO PROBLEM.

GEOMETRIC INTERPRETATIONS

LET'S HAVE A LOOK AT HOW TO EXPRESS POINTS, LINES, AND SPACES WITH VECTORS.

THE NOTATION MIGHT LOOK A BIT WEIRD AT FIRST, BUT YOU'LL GET USED TO IT.

A PLANE

AND THE x_1x_2 PLANE R^2 CAN BE EXPRESSED AS:

$$\left\{ c_1\begin{pmatrix}1\\0\end{pmatrix}+c_2\begin{pmatrix}0\\1\end{pmatrix} \middle| \begin{array}{l}c_1, c_2 \text{ are arbitrary}\\ \text{real numbers}\end{array}\right\}$$

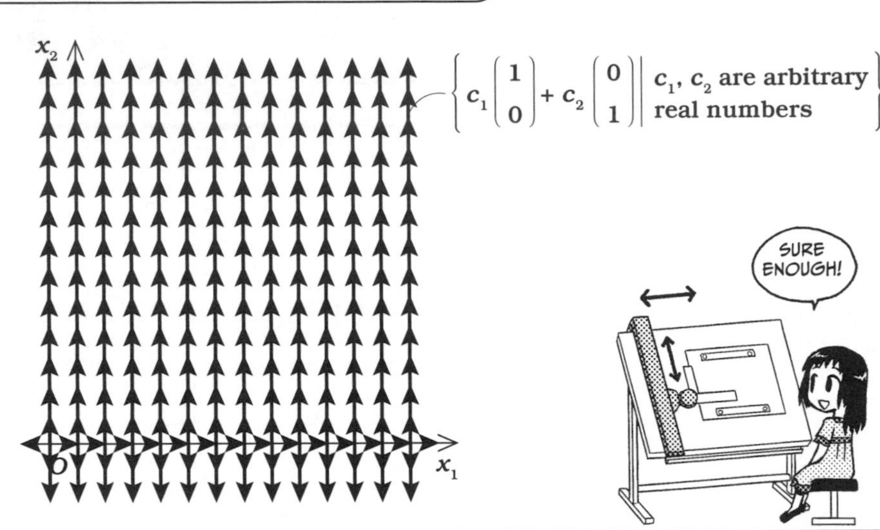

ANOTHER PLANE

IT CAN ALSO BE WRITTEN ANOTHER WAY:

$$\left\{ c_1\begin{pmatrix}3\\1\end{pmatrix}+c_2\begin{pmatrix}1\\2\end{pmatrix} \middle| \begin{array}{l}c_1, c_2 \text{ are arbitrary}\\ \text{real numbers}\end{array}\right\}$$

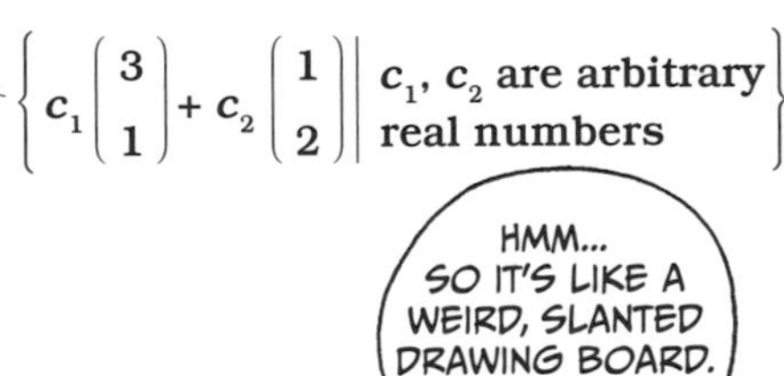

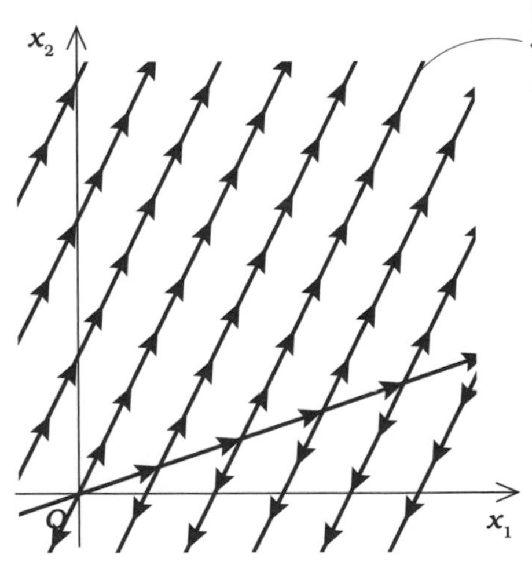

A VECTOR SPACE

THE THREE-DIMENSIONAL SPACE R^3 IS THE NATURAL NEXT STEP. IT IS SPANNED BY x_1, x_2, AND x_3 LIKE THIS:

$$\left\{ c_1 \begin{pmatrix} 1 \\ 0 \\ 0 \end{pmatrix} + c_2 \begin{pmatrix} 0 \\ 1 \\ 0 \end{pmatrix} + c_3 \begin{pmatrix} 0 \\ 0 \\ 1 \end{pmatrix} \middle| \; c_1, c_2, c_3 \text{ are arbitrary real numbers} \right\}$$

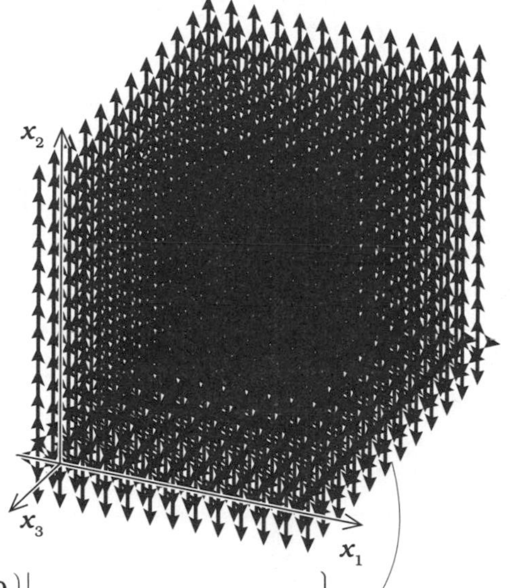

$$\left\{ c_1 \begin{pmatrix} 1 \\ 0 \\ 0 \end{pmatrix} + c_2 \begin{pmatrix} 0 \\ 1 \\ 0 \end{pmatrix} + c_3 \begin{pmatrix} 0 \\ 0 \\ 1 \end{pmatrix} \middle| \; c_1, c_2, c_3 \text{ are arbitrary real numbers} \right\}$$

ANOTHER VECTOR SPACE

NOW TRY TO IMAGINE THE n-DIMENSIONAL SPACE R^n, SPANNED BY x_1, x_2, ..., x_n:

$$\left\{ c_1 \begin{pmatrix} 1 \\ 0 \\ \vdots \\ 0 \end{pmatrix} + c_2 \begin{pmatrix} 0 \\ 1 \\ \vdots \\ 0 \end{pmatrix} + \dots + c_n \begin{pmatrix} 0 \\ 0 \\ \vdots \\ 1 \end{pmatrix} \middle| \; c_1, c_2, \dots, c_n \text{ are arbitrary real numbers} \right\}$$

I UNDERSTAND THE FORMULA, BUT THIS ONE'S A LITTLE HARDER TO VISUALIZE...

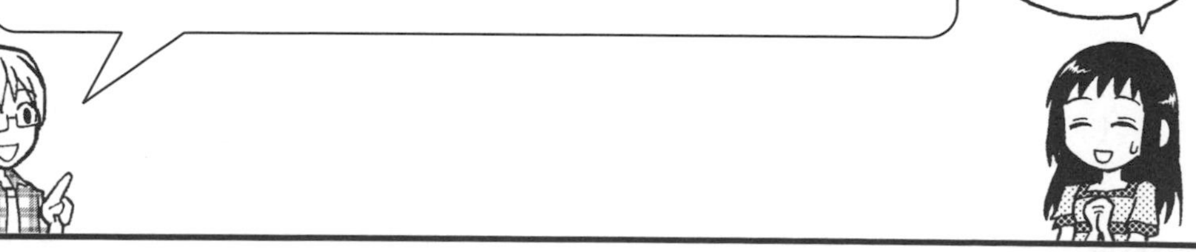

WHOO, I'M BEAT!

LET'S TAKE A BREAK, THEN.
GOOD IDEA!

BY THE WAY, REIJI...

WHY DID YOU DECIDE TO JOIN THE KARATE CLUB, ANYWAY?
EH?
UM, WELL...
NO SPECIAL REASON REALLY...
HEY!
WE'D BETTER GET BACK TO WORK!
?

6
MORE VECTORS

LINEAR INDEPENDENCE

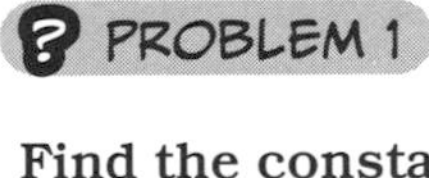

Find the constants c_1 and c_2 satisfying this equation:

$$\begin{pmatrix} 0 \\ 0 \end{pmatrix} = c_1 \begin{pmatrix} 1 \\ 0 \end{pmatrix} + c_2 \begin{pmatrix} 0 \\ 1 \end{pmatrix}$$

PROBLEM 2

Find the constants c_1 and c_2 satisfying this equation:

$$\begin{pmatrix} 0 \\ 0 \end{pmatrix} = c_1 \begin{pmatrix} 3 \\ 1 \end{pmatrix} + c_2 \begin{pmatrix} 1 \\ 2 \end{pmatrix}$$

PROBLEM 3

Find the constants c_1, c_2, c_3, and c_4 satisfying this equation:

$$\begin{pmatrix} 0 \\ 0 \end{pmatrix} = c_1 \begin{pmatrix} 1 \\ 0 \end{pmatrix} + c_2 \begin{pmatrix} 0 \\ 1 \end{pmatrix} + c_3 \begin{pmatrix} 3 \\ 1 \end{pmatrix} + c_4 \begin{pmatrix} 1 \\ 2 \end{pmatrix}$$

$\begin{cases} c_1 = 0 \\ c_2 = 0 \\ c_3 = 0 \\ c_4 = 0 \end{cases}$
AGAIN?

NOT QUITE.
?

$\begin{cases} c_1 = 0 \\ c_2 = 0 \\ c_3 = 0 \\ c_4 = 0 \end{cases}$
ISN'T WRONG, BUT...

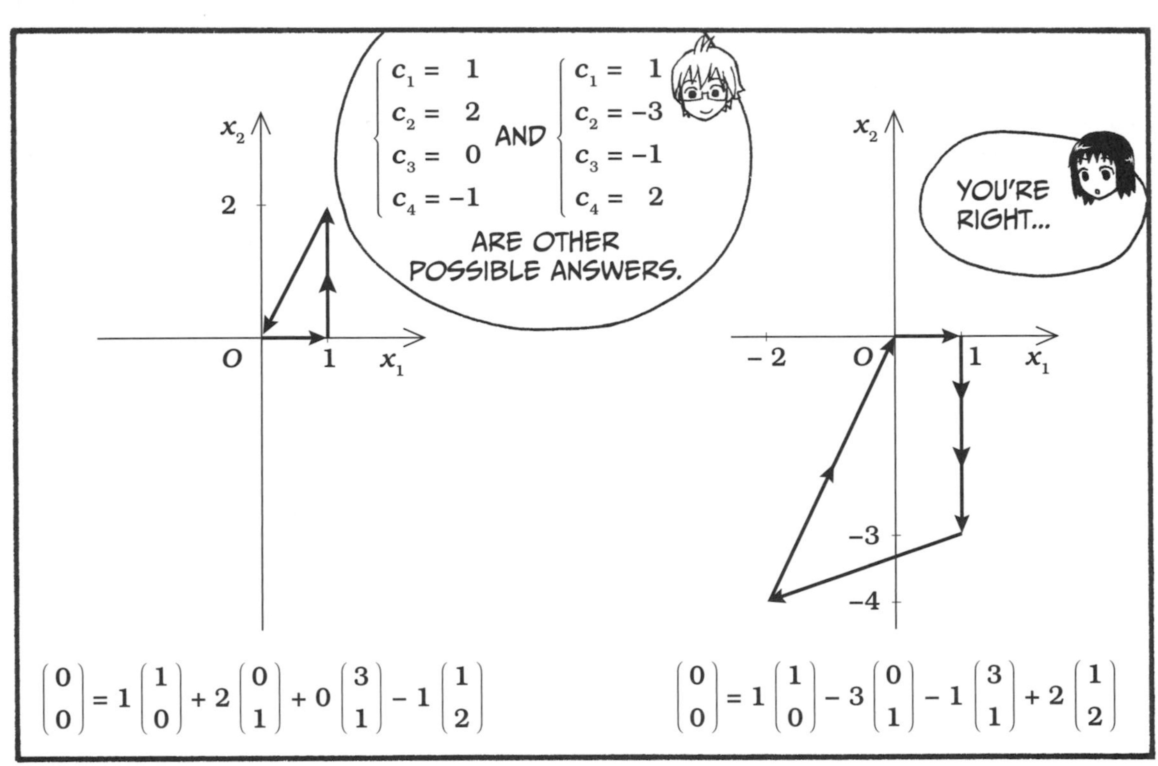
$\begin{cases} c_1 = 1 \\ c_2 = 2 \\ c_3 = 0 \\ c_4 = -1 \end{cases}$ AND $\begin{cases} c_1 = 1 \\ c_2 = -3 \\ c_3 = -1 \\ c_4 = 2 \end{cases}$
ARE OTHER POSSIBLE ANSWERS.
YOU'RE RIGHT...
x_2
2
O
1
x_1
x_2
− 2
O
1
x_1
−3
−4
$\begin{pmatrix} 0 \\ 0 \end{pmatrix} = 1 \begin{pmatrix} 1 \\ 0 \end{pmatrix} + 2 \begin{pmatrix} 0 \\ 1 \end{pmatrix} + 0 \begin{pmatrix} 3 \\ 1 \end{pmatrix} - 1 \begin{pmatrix} 1 \\ 2 \end{pmatrix}$
$\begin{pmatrix} 0 \\ 0 \end{pmatrix} = 1 \begin{pmatrix} 1 \\ 0 \end{pmatrix} - 3 \begin{pmatrix} 0 \\ 1 \end{pmatrix} - 1 \begin{pmatrix} 3 \\ 1 \end{pmatrix} + 2 \begin{pmatrix} 1 \\ 2 \end{pmatrix}$

TRY TO KEEP THIS IN MIND WHILE WE MOVE ON TO THE MAIN PROBLEM.

AS LONG AS THERE IS ONLY ONE UNIQUE SOLUTION

$$\begin{cases} c_1 = 0 \\ c_2 = 0 \\ \vdots \\ c_n = 0 \end{cases}$$

TO PROBLEMS SUCH AS THE FIRST OR SECOND EXAMPLES:

$$\begin{pmatrix} 0 \\ 0 \\ \vdots \\ 0 \end{pmatrix} = c_1 \begin{pmatrix} a_{11} \\ a_{21} \\ \vdots \\ a_{m1} \end{pmatrix} + c_2 \begin{pmatrix} a_{12} \\ a_{22} \\ \vdots \\ a_{m2} \end{pmatrix} + \ldots + c_n \begin{pmatrix} a_{1n} \\ a_{2n} \\ \vdots \\ a_{mn} \end{pmatrix}$$

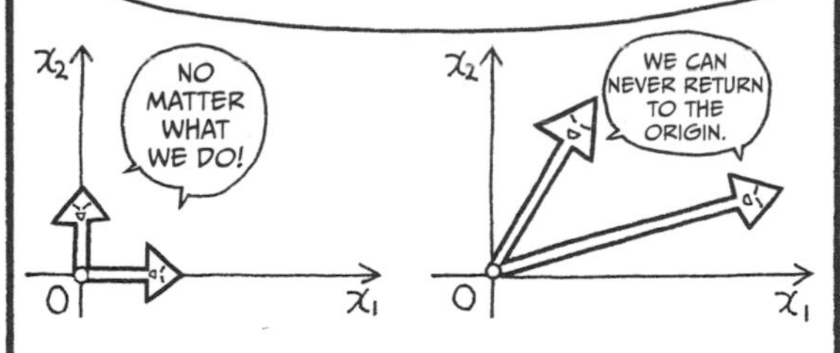

LINEAR INDEPENDENCE

WE SAY THAT ITS VECTORS

$$\begin{pmatrix} a_{11} \\ a_{21} \\ \vdots \\ a_{m1} \end{pmatrix}, \begin{pmatrix} a_{12} \\ a_{22} \\ \vdots \\ a_{m2} \end{pmatrix}, \text{AND} \begin{pmatrix} a_{1n} \\ a_{2n} \\ \vdots \\ a_{mn} \end{pmatrix}$$

ARE *LINEARLY INDEPENDENT.*

AS FOR PROBLEMS LIKE THE THIRD EXAMPLE, WHERE THERE ARE SOLUTIONS OTHER THAN

$$\begin{cases} c_1 = 0 \\ c_2 = 0 \\ \vdots \\ c_n = 0 \end{cases}$$

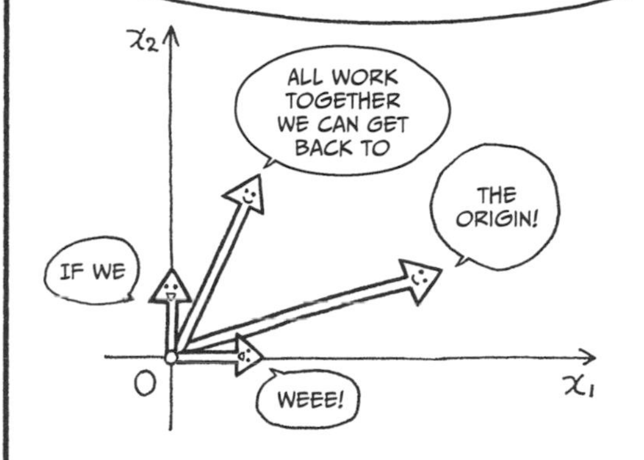

LINEAR DEPENDENCE

THEIR VECTORS

$$\begin{pmatrix} a_{11} \\ a_{21} \\ \vdots \\ a_{m1} \end{pmatrix}, \begin{pmatrix} a_{12} \\ a_{22} \\ \vdots \\ a_{m2} \end{pmatrix}, \text{AND} \begin{pmatrix} a_{1n} \\ a_{2n} \\ \vdots \\ a_{mn} \end{pmatrix}$$

ARE CALLED *LINEARLY DEPENDENT.*

HERE ARE SOME EXAMPLES. LET'S LOOK AT LINEAR INDEPENDENCE FIRST.

EXAMPLE 1

The vectors $\begin{pmatrix} 1 \\ 0 \\ 0 \end{pmatrix}$, $\begin{pmatrix} 0 \\ 1 \\ 0 \end{pmatrix}$, and $\begin{pmatrix} 0 \\ 0 \\ 1 \end{pmatrix}$

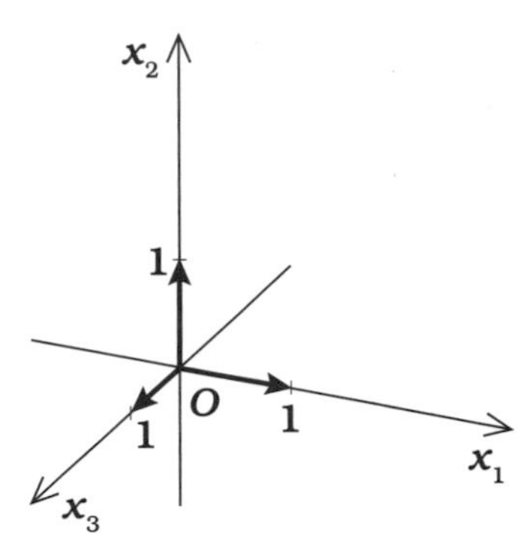

give us the equation $\begin{pmatrix} 0 \\ 0 \\ 0 \end{pmatrix} = c_1 \begin{pmatrix} 1 \\ 0 \\ 0 \end{pmatrix} + c_2 \begin{pmatrix} 0 \\ 1 \\ 0 \end{pmatrix} + c_3 \begin{pmatrix} 0 \\ 0 \\ 1 \end{pmatrix}$

which has the unique solution $\begin{cases} c_1 = 0 \\ c_2 = 0 \\ c_3 = 0 \end{cases}$

The vectors are therefore linearly independent.

EXAMPLE 2

The vectors $\begin{pmatrix} 1 \\ 0 \\ 0 \end{pmatrix}$ and $\begin{pmatrix} 0 \\ 1 \\ 0 \end{pmatrix}$

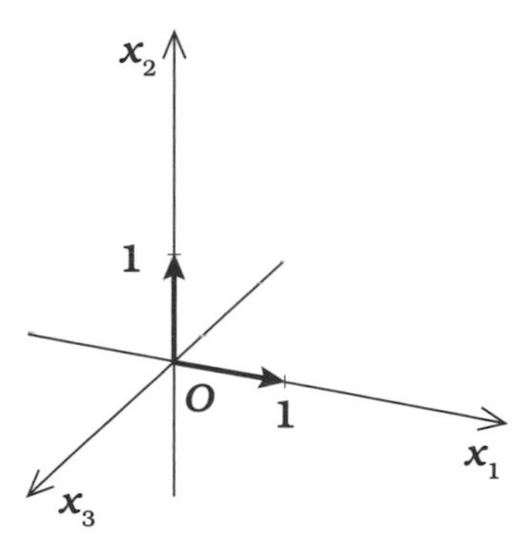

give us the equation $\begin{pmatrix} 0 \\ 0 \\ 0 \end{pmatrix} = c_1 \begin{pmatrix} 1 \\ 0 \\ 0 \end{pmatrix} + c_2 \begin{pmatrix} 0 \\ 1 \\ 0 \end{pmatrix}$

which has the unique solution $\begin{cases} c_1 = 0 \\ c_2 = 0 \end{cases}$

These vectors are therefore also linearly independent.

AND NOW WE'LL LOOK AT LINEAR DEPENDENCE.

EXAMPLE 1

The vectors $\begin{pmatrix}1\\0\\0\end{pmatrix}$, $\begin{pmatrix}0\\1\\0\end{pmatrix}$, and $\begin{pmatrix}3\\1\\0\end{pmatrix}$

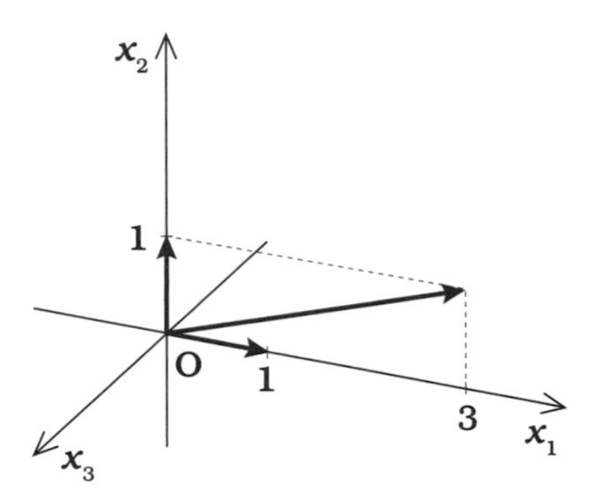

give us the equation $\begin{pmatrix}0\\0\\0\end{pmatrix} = c_1\begin{pmatrix}1\\0\\0\end{pmatrix} + c_2\begin{pmatrix}0\\1\\0\end{pmatrix} + c_3\begin{pmatrix}3\\1\\0\end{pmatrix}$

which has several solutions, for example $\begin{cases}c_1 = 0\\c_2 = 0\\c_3 = 0\end{cases}$ and $\begin{cases}c_1 = 3\\c_2 = 1\\c_3 = -1\end{cases}$

This means that the vectors are linearly dependent.

EXAMPLE 2

Suppose we have the vectors $\begin{pmatrix} 1 \\ 0 \\ 0 \end{pmatrix}$, $\begin{pmatrix} 0 \\ 1 \\ 0 \end{pmatrix}$, $\begin{pmatrix} 0 \\ 0 \\ 1 \end{pmatrix}$, and $\begin{pmatrix} a_1 \\ a_2 \\ a_3 \end{pmatrix}$

as well as the equation $\begin{pmatrix} 0 \\ 0 \\ 0 \end{pmatrix} = c_1 \begin{pmatrix} 1 \\ 0 \\ 0 \end{pmatrix} + c_2 \begin{pmatrix} 0 \\ 1 \\ 0 \end{pmatrix} + c_3 \begin{pmatrix} 0 \\ 0 \\ 1 \end{pmatrix} + c_4 \begin{pmatrix} a_1 \\ a_2 \\ a_3 \end{pmatrix}$

The vectors are linearly dependent because there are several solutions to the system—

for example, $\begin{cases} c_1 = 0 \\ c_2 = 0 \\ c_3 = 0 \\ c_4 = 0 \end{cases}$ and $\begin{cases} c_1 = a_1 \\ c_2 = a_2 \\ c_3 = a_3 \\ c_4 = -1 \end{cases}$

The vectors $\begin{pmatrix} 1 \\ 0 \\ \vdots \\ 0 \end{pmatrix}$, $\begin{pmatrix} 0 \\ 1 \\ \vdots \\ 0 \end{pmatrix}$, $\begin{pmatrix} 0 \\ 0 \\ \vdots \\ 1 \end{pmatrix}$, and $\begin{pmatrix} a_1 \\ a_2 \\ \vdots \\ a_m \end{pmatrix}$

are similarly linearly dependent because there are several solutions to the equation

$$\begin{pmatrix} 0 \\ 0 \\ \vdots \\ 0 \end{pmatrix} = c_1 \begin{pmatrix} 1 \\ 0 \\ \vdots \\ 0 \end{pmatrix} + c_2 \begin{pmatrix} 0 \\ 1 \\ \vdots \\ 0 \end{pmatrix} + \ldots + c_m \begin{pmatrix} 0 \\ 0 \\ \vdots \\ 1 \end{pmatrix} + c_{m+1} \begin{pmatrix} a_1 \\ a_2 \\ \vdots \\ a_m \end{pmatrix}$$

Among them is $\begin{cases} c_1 = 0 \\ c_2 = 0 \\ \quad\vdots \\ c_m = 0 \\ c_{m+1} = 0 \end{cases}$ but also $\begin{cases} c_1 = a_1 \\ c_2 = a_2 \\ \quad\vdots \\ c_m = a_m \\ c_{m+1} = -1 \end{cases}$

PROBLEM 4

Find the constants c_1 and c_2 satisfying this equation:

$$\begin{pmatrix} 7 \\ 4 \end{pmatrix} = c_1 \begin{pmatrix} 1 \\ 0 \end{pmatrix} + c_2 \begin{pmatrix} 0 \\ 1 \end{pmatrix}$$

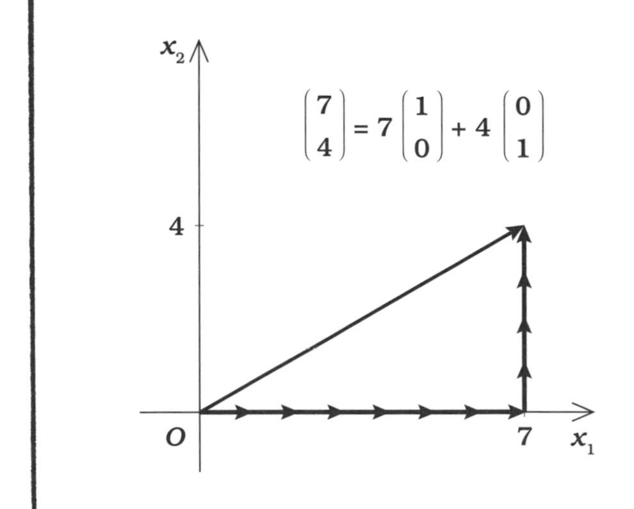

$$\begin{cases} c_1 = 7 \\ c_2 = 4 \end{cases}$$

SHOULD WORK.

HERE'S THE SECOND ONE.

PROBLEM 5

Find the constants c_1 and c_2 satisfying this equation:

$$\begin{pmatrix}7\\4\end{pmatrix} = c_1\begin{pmatrix}3\\1\end{pmatrix} + c_2\begin{pmatrix}1\\2\end{pmatrix}$$

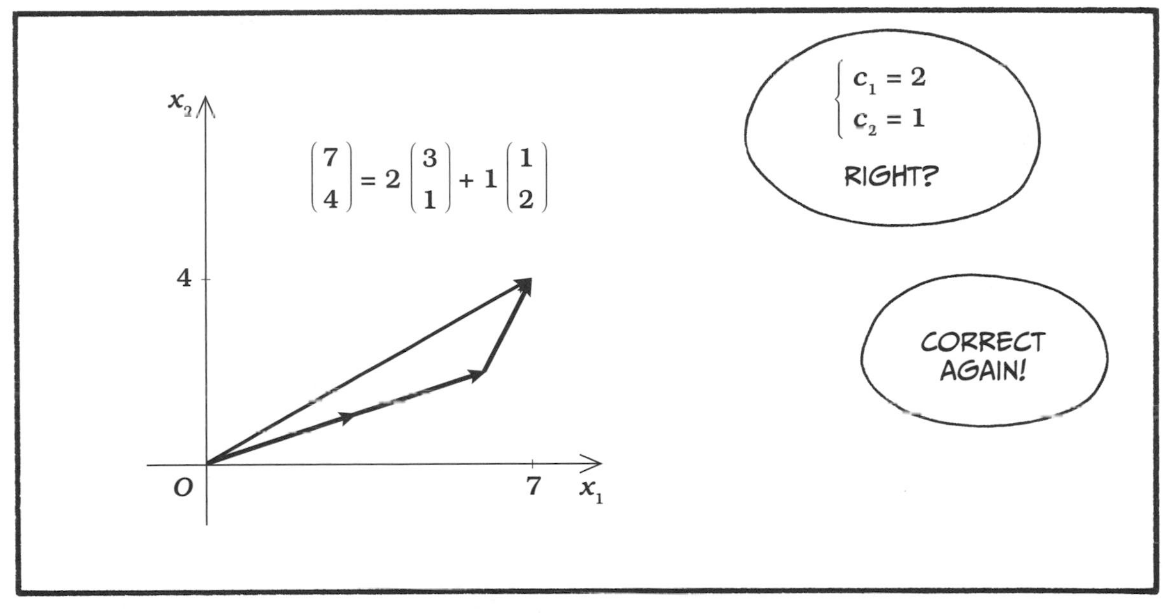

LAST ONE.

PROBLEM 6

Find the constants c_1, c_2, c_3, and c_4 satisfying this equation:

$$\begin{pmatrix} 7 \\ 4 \end{pmatrix} = c_1 \begin{pmatrix} 1 \\ 0 \end{pmatrix} + c_2 \begin{pmatrix} 0 \\ 1 \end{pmatrix} + c_3 \begin{pmatrix} 3 \\ 1 \end{pmatrix} + c_4 \begin{pmatrix} 1 \\ 2 \end{pmatrix}$$

LINEAR DEPENDENCE AND INDEPENDENCE ARE CLOSELY RELATED TO THE CONCEPT OF A *BASIS*. HAVE A LOOK AT THE FOLLOWING EQUATION:

$$\begin{pmatrix} y_1 \\ y_2 \\ \vdots \\ y_m \end{pmatrix} = c_1 \begin{pmatrix} a_{11} \\ a_{21} \\ \vdots \\ a_{m1} \end{pmatrix} + c_2 \begin{pmatrix} a_{12} \\ a_{22} \\ \vdots \\ a_{m2} \end{pmatrix} + \dots + c_n \begin{pmatrix} a_{1n} \\ a_{2n} \\ \vdots \\ a_{mn} \end{pmatrix}$$

WHERE THE LEFT SIDE OF THE EQUATION IS AN ARBITRARY VECTOR IN R^m AND THE RIGHT SIDE IS A NUMBER OF n VECTORS OF THE SAME DIMENSION, AS WELL AS THEIR COEFFICIENTS.

IF THERE'S ONLY ONE SOLUTION
$c_1 = c_2 = \dots = c_n = 0$
TO THE EQUATION, THEN OUR VECTORS

$$\left\{ \begin{pmatrix} a_{11} \\ a_{21} \\ \vdots \\ a_{m1} \end{pmatrix}, \begin{pmatrix} a_{12} \\ a_{22} \\ \vdots \\ a_{m2} \end{pmatrix}, \dots, \begin{pmatrix} a_{1n} \\ a_{2n} \\ \vdots \\ a_{mn} \end{pmatrix} \right\}$$

MAKE UP A BASIS FOR R^n.

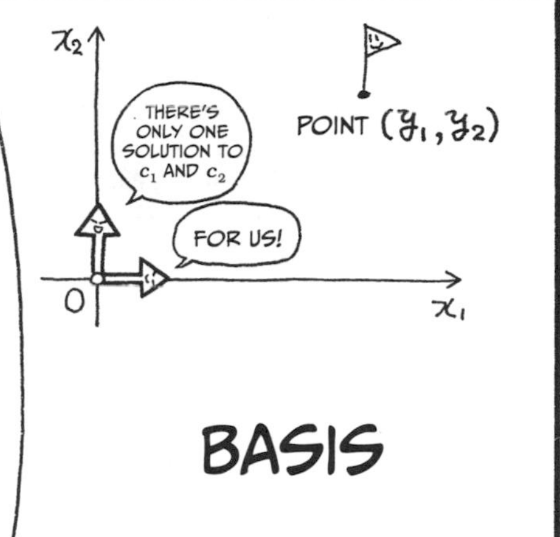

BASIS

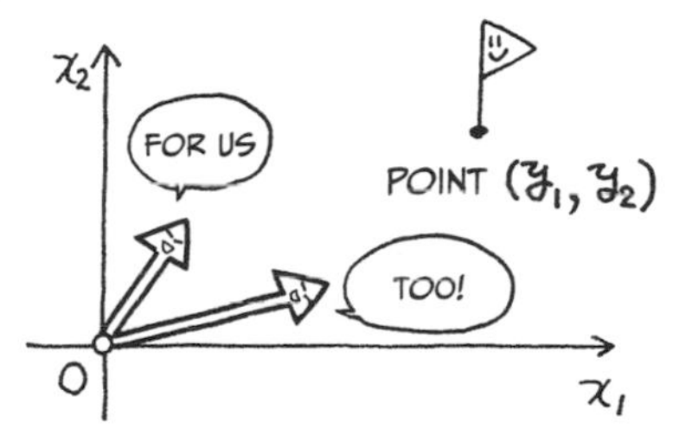

ALL THESE VECTOR SETS MAKE UP BASES FOR THEIR GRAPHS.

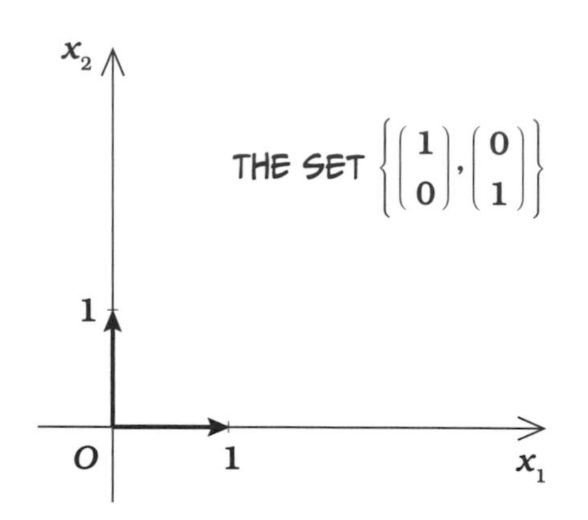

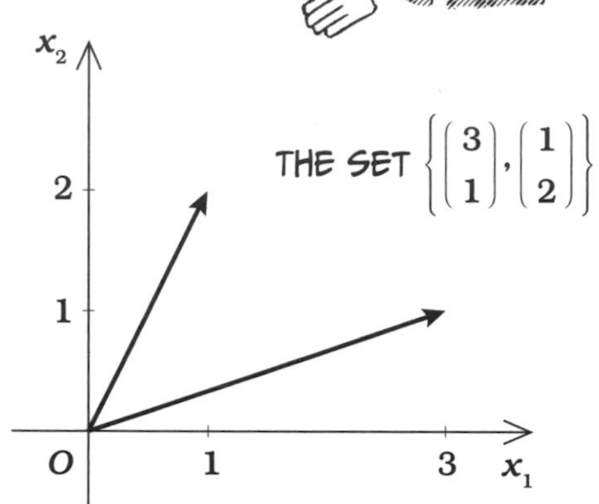

THE SET $\left\{\begin{pmatrix}1\\0\\0\end{pmatrix},\begin{pmatrix}0\\1\\0\end{pmatrix},\begin{pmatrix}0\\0\\1\end{pmatrix}\right\}$

x_2 1 O 1 1 x_1 x_3

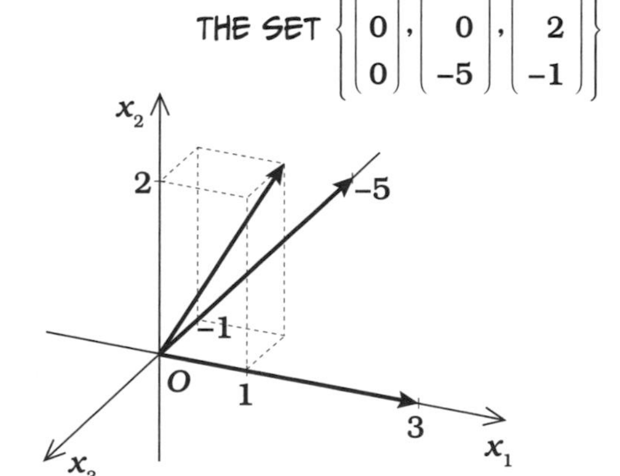

IN OTHER WORDS, A *BASIS* IS A MINIMAL SET OF VECTORS NEEDED TO EXPRESS AN ARBITRARY VECTOR IN R^m. ANOTHER IMPORTANT FEATURE OF BASES IS THAT THEY'RE ALL LINEARLY INDEPENDENT.

THE VECTORS OF THE FOLLOWING SET DO *NOT* FORM A BASIS.

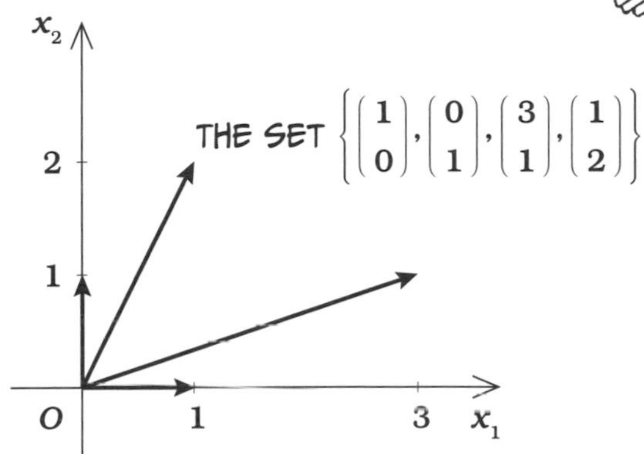

TO UNDERSTAND WHY THEY DON'T FORM A BASIS, HAVE A LOOK AT THE FOLLOWING EQUATION:

$$\begin{pmatrix} y_1 \\ y_2 \end{pmatrix} = c_1 \begin{pmatrix} 1 \\ 0 \end{pmatrix} + c_2 \begin{pmatrix} 0 \\ 1 \end{pmatrix} + c_3 \begin{pmatrix} 3 \\ 1 \end{pmatrix} + c_4 \begin{pmatrix} 1 \\ 2 \end{pmatrix}$$

WHERE $\begin{pmatrix} y_1 \\ y_2 \end{pmatrix}$ IS AN ARBITRARY VECTOR IN R^2.

$\begin{pmatrix} y_1 \\ y_2 \end{pmatrix}$ CAN BE FORMED IN MANY DIFFERENT WAYS (USING DIFFERENT CHOICES FOR c_1, c_2, c_3, AND c_4).

BECAUSE OF THIS, THE SET DOES NOT FORM "A MINIMAL SET OF VECTORS NEEDED TO EXPRESS AN ARBITRARY VECTOR IN R^m."

NEITHER OF THE TWO VECTOR SETS BELOW IS ABLE TO DESCRIBE THE VECTOR $\begin{pmatrix}0\\0\\1\end{pmatrix}$, AND IF THEY CAN'T DESCRIBE THAT VECTOR, THEN THERE'S NO WAY THAT THEY COULD DESCRIBE "AN ARBITRARY VECTOR IN R^3." BECAUSE OF THIS, THEY'RE NOT BASES.

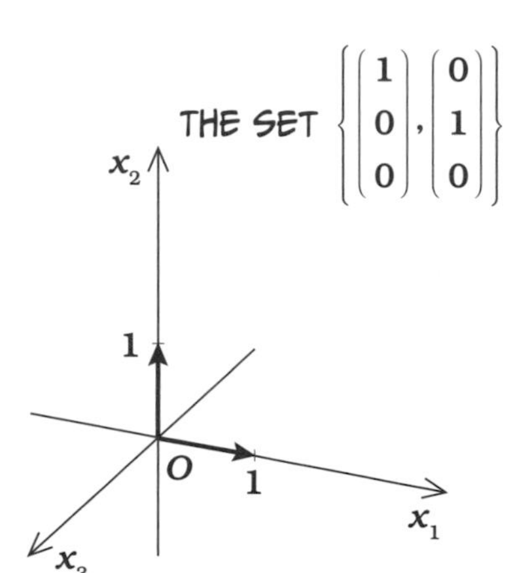

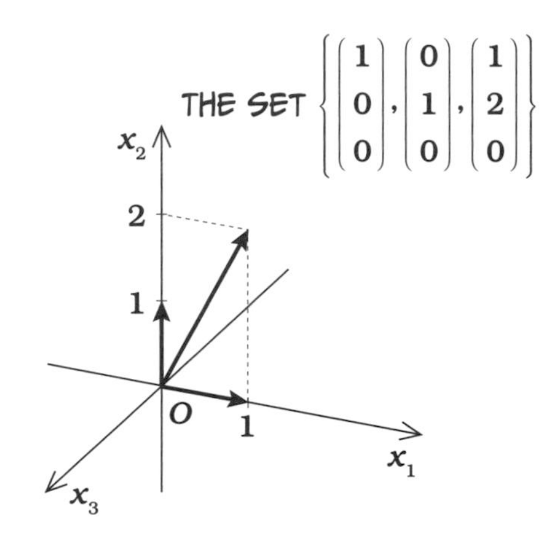

JUST BECAUSE A SET OF VECTORS IS LINEARLY INDEPENDENT DOESN'T MEAN THAT IT FORMS A BASIS. FOR INSTANCE, THE SET $\left\{\begin{pmatrix}1\\0\\0\end{pmatrix}, \begin{pmatrix}0\\1\\0\end{pmatrix}, \begin{pmatrix}0\\0\\1\end{pmatrix}\right\}$ FORMS A BASIS, WHILE THE SET $\left\{\begin{pmatrix}1\\0\\0\end{pmatrix}, \begin{pmatrix}0\\1\\0\end{pmatrix}\right\}$ DOES NOT, EVEN THOUGH THEY'RE BOTH LINEARLY INDEPENDENT.

LINEAR INDEPENDENCE

We say that a set of vectors $\left\{\begin{pmatrix} a_{11} \\ a_{21} \\ \vdots \\ a_{m1} \end{pmatrix}, \begin{pmatrix} a_{12} \\ a_{22} \\ \vdots \\ a_{m2} \end{pmatrix}, \dots, \begin{pmatrix} a_{1n} \\ a_{2n} \\ \vdots \\ a_{mn} \end{pmatrix}\right\}$ is linearly independent

if there's only one solution $\begin{cases} c_1 = 0 \\ c_2 = 0 \\ \vdots \\ c_n = 0 \end{cases}$

to the equation $\begin{pmatrix} 0 \\ 0 \\ \vdots \\ 0 \end{pmatrix} = c_1 \begin{pmatrix} a_{11} \\ a_{21} \\ \vdots \\ a_{m1} \end{pmatrix} + c_2 \begin{pmatrix} a_{12} \\ a_{22} \\ \vdots \\ a_{m2} \end{pmatrix} + \dots + c_n \begin{pmatrix} a_{1n} \\ a_{2n} \\ \vdots \\ a_{mn} \end{pmatrix}$

where the left side is the zero vector of R^m.

BASES

A set of vectors $\left\{\begin{pmatrix} a_{11} \\ a_{21} \\ \vdots \\ a_{m1} \end{pmatrix}, \begin{pmatrix} a_{12} \\ a_{22} \\ \vdots \\ a_{m2} \end{pmatrix}, \dots, \begin{pmatrix} a_{1n} \\ a_{2n} \\ \vdots \\ a_{mn} \end{pmatrix}\right\}$ forms a basis if there's only

one solution to the equation $\begin{pmatrix} y_1 \\ y_2 \\ \vdots \\ y_m \end{pmatrix} = c_1 \begin{pmatrix} a_{11} \\ a_{21} \\ \vdots \\ a_{m1} \end{pmatrix} + c_2 \begin{pmatrix} a_{12} \\ a_{22} \\ \vdots \\ a_{m2} \end{pmatrix} + \dots + c_n \begin{pmatrix} a_{1n} \\ a_{2n} \\ \vdots \\ a_{mn} \end{pmatrix}$

where the left side is an arbitrary vector $\begin{pmatrix} y_1 \\ y_2 \\ \vdots \\ y_m \end{pmatrix}$ in R^m. And once again, a basis

is a minimal set of vectors needed to express an arbitrary vector in R^m.

SO...

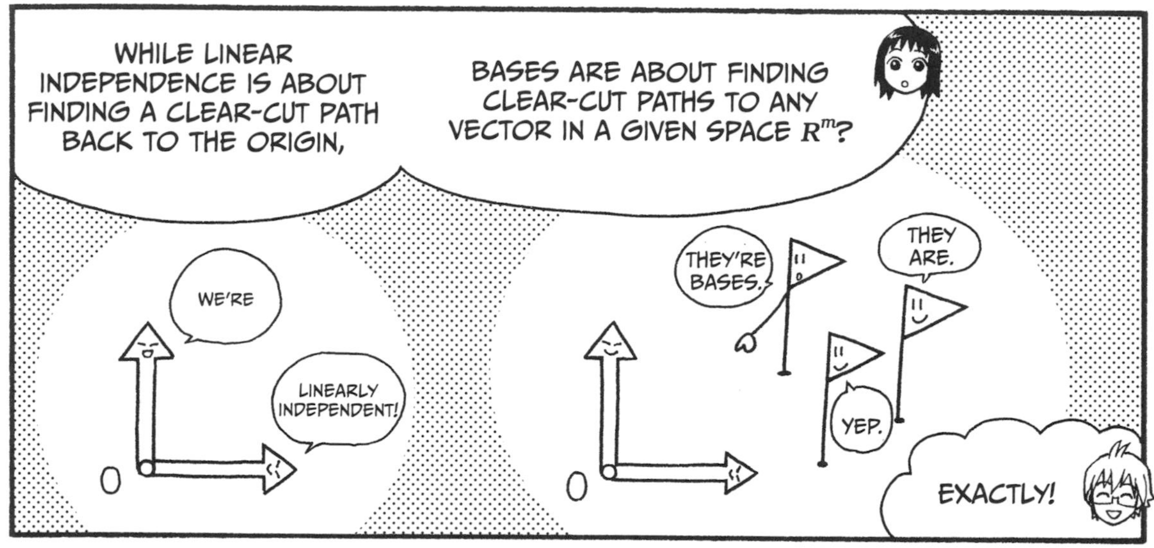
WHILE LINEAR INDEPENDENCE IS ABOUT FINDING A CLEAR-CUT PATH BACK TO THE ORIGIN,
BASES ARE ABOUT FINDING CLEAR-CUT PATHS TO ANY VECTOR IN A GIVEN SPACE R^m?
WE'RE
LINEARLY INDEPENDENT!
O
THEY'RE BASES.
THEY ARE.
YEP.
O
EXACTLY!

NOT A LOT OF PEOPLE ARE ABLE TO GRASP THE DIFFERENCE BETWEEN THE TWO THAT FAST! I MUST SAY I'M IMPRESSED!
NO BIG DEAL!

THAT'S ALL FOR TOD–
AH, WAIT A SEC!

DIMENSION

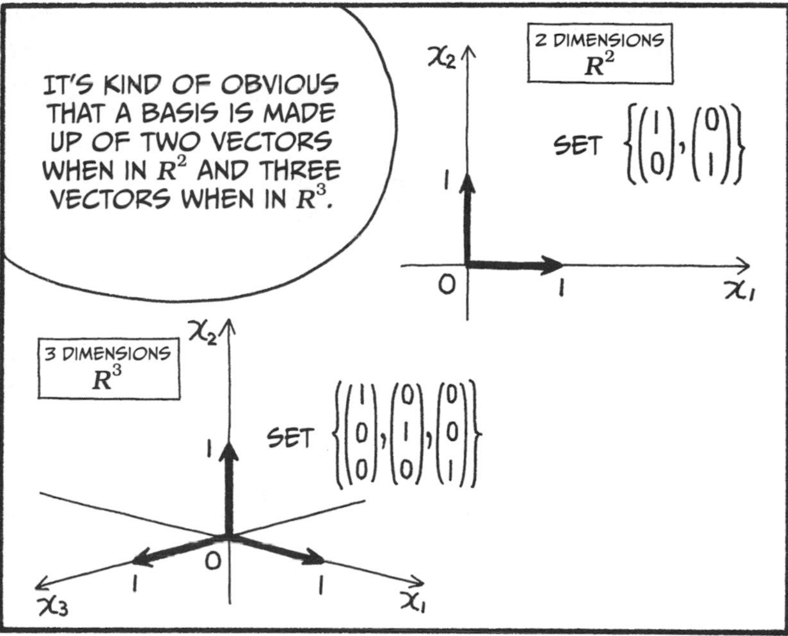

BUT WHY IS IT THAT THE BASIS OF AN m-DIMENSIONAL SPACE CONSISTS OF n VECTORS AND NOT m?

$$\left\{\begin{pmatrix}a_{11}\\a_{21}\\\vdots\\a_{m1}\end{pmatrix}, \begin{pmatrix}a_{12}\\a_{22}\\\vdots\\a_{m2}\end{pmatrix}, \ldots, \begin{pmatrix}a_{1n}\\a_{2n}\\\vdots\\a_{mn}\end{pmatrix}\right\}$$

OH, WOW... I DIDN'T THINK YOU'D NOTICE...

YOU SURE?
I—
I THINK SO.

IT'S ACTUALLY NOT *THAT* HARD—JUST A LITTLE ABSTRACT.
LET'S HAVE A LOOK, SINCE YOU ASKED AND ALL.
O-OKAY...

BUT FIRST WE HAVE TO TACKLE ANOTHER NEW CONCEPT: SUBSPACES.
SUBSPACE
SO LET'S TALK ABOUT THEM.

SUBSPACES

IT'S KINDA LIKE THIS.
R^m
W

SO IT'S ANOTHER WORD FOR SUBSET?
NO, NOT QUITE. LET ME TRY AGAIN.

WHAT IS A SUBSPACE?

Let c be an arbitrary real number and W be a nonempty subset of R^m satisfying these two conditions:

❶ An element in W multiplied by c is still an element in W. (Closed under scalar multiplication.)

$$\text{If } \begin{pmatrix} a_{1i} \\ a_{2i} \\ \vdots \\ a_{mi} \end{pmatrix} \in W, \text{ then } c\begin{pmatrix} a_{1i} \\ a_{2i} \\ \vdots \\ a_{mi} \end{pmatrix} \in W$$

❷ The sum of two arbitrary elements in W is still an element in W. (Closed under addition.)

$$\text{If } \begin{pmatrix} a_{1i} \\ a_{2i} \\ \vdots \\ a_{mi} \end{pmatrix} \in W \text{ and } \begin{pmatrix} a_{1j} \\ a_{2j} \\ \vdots \\ a_{mj} \end{pmatrix} \in W, \text{ then } \begin{pmatrix} a_{1i} \\ a_{2i} \\ \vdots \\ a_{mi} \end{pmatrix} + \begin{pmatrix} a_{1j} \\ a_{2j} \\ \vdots \\ a_{mj} \end{pmatrix} \in W$$

If both of these conditions hold, then W is a subspace of R^m.

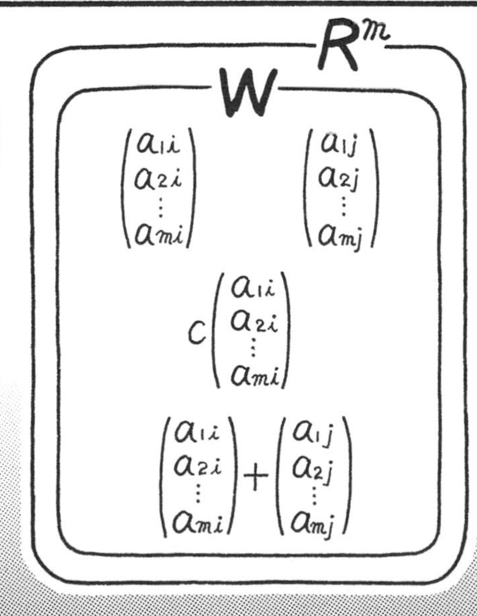

IT'S PRETTY ABSTRACT, SO YOU MIGHT HAVE TO READ IT A FEW TIMES BEFORE IT STARTS TO SINK IN.

ANOTHER, MORE CONCRETE WAY TO LOOK AT ONE-DIMENSIONAL SUBSPACES IS AS LINES THROUGH THE ORIGIN. TWO-DIMENSIONAL SUBSPACES ARE SIMILARLY PLANES THROUGH THE ORIGIN. OTHER SUBSPACES CAN ALSO BE VISUALIZED, BUT NOT AS EASILY.

I MADE SOME EXAMPLES OF SPACES THAT ARE SUBSPACES—AND OF SOME THAT ARE NOT. HAVE A LOOK!

THIS IS A SUBSPACE

Let's have a look at the subspace in R^3 defined by the set

$$\left\{ \begin{pmatrix} \alpha \\ 0 \\ 0 \end{pmatrix} \middle| \begin{array}{l} \alpha \text{ is an} \\ \text{arbitrary} \\ \text{real number} \end{array} \right\}$$, in other words, the *x*-axis.

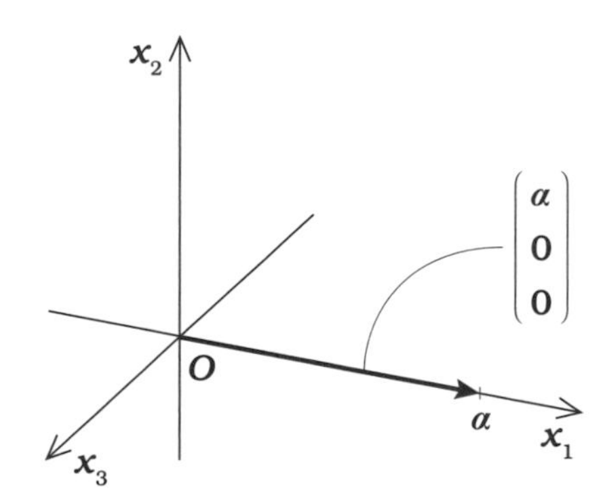

If it really is a subspace, it should satisfy the two conditions we talked about before.

❶ $$c\begin{pmatrix} \alpha_1 \\ 0 \\ 0 \end{pmatrix} = \begin{pmatrix} c\alpha_1 \\ 0 \\ 0 \end{pmatrix} \in \left\{ \begin{pmatrix} \alpha \\ 0 \\ 0 \end{pmatrix} \middle| \begin{array}{l} \alpha \text{ is an} \\ \text{arbitrary} \\ \text{real number} \end{array} \right\}$$

❷ $$\begin{pmatrix} \alpha_1 \\ 0 \\ 0 \end{pmatrix} + \begin{pmatrix} \alpha_2 \\ 0 \\ 0 \end{pmatrix} = \begin{pmatrix} \alpha_1+\alpha_2 \\ 0 \\ 0 \end{pmatrix} \in \left\{ \begin{pmatrix} \alpha \\ 0 \\ 0 \end{pmatrix} \middle| \begin{array}{l} \alpha \text{ is an} \\ \text{arbitrary} \\ \text{real number} \end{array} \right\}$$

It seems like they do! This means it actually is a subspace.

THIS IS NOT A SUBSPACE

The set $\left\{ \begin{pmatrix} \alpha \\ \alpha^2 \\ 0 \end{pmatrix} \middle| \ \alpha \text{ is an arbitrary real number} \right\}$ is not a subspace of R^3.

Let's use our conditions to see why:

❶ $$c\begin{pmatrix} \alpha_1 \\ \alpha_1^{\ 2} \\ 0 \end{pmatrix} = \begin{pmatrix} c\alpha_1 \\ c\alpha_1^{\ 2} \\ 0 \end{pmatrix} \neq \begin{pmatrix} c\alpha_1 \\ (c\alpha_1)^2 \\ 0 \end{pmatrix} \in \left\{ \begin{pmatrix} \alpha \\ \alpha^2 \\ 0 \end{pmatrix} \middle| \ \alpha \text{ is an arbitrary real number} \right\}$$

❷ $$\begin{pmatrix} \alpha_1 \\ \alpha_1^{\ 2} \\ 0 \end{pmatrix} + \begin{pmatrix} \alpha_2 \\ \alpha_2^{\ 2} \\ 0 \end{pmatrix} = \begin{pmatrix} \alpha_1+\alpha_2 \\ \alpha_1^{\ 2}+\alpha_2^{\ 2} \\ 0 \end{pmatrix} \neq \begin{pmatrix} \alpha_1+\alpha_2 \\ (\alpha_1+\alpha_2)^2 \\ 0 \end{pmatrix} \in \left\{ \begin{pmatrix} \alpha \\ \alpha^2 \\ 0 \end{pmatrix} \middle| \ \alpha \text{ is an arbitrary real number} \right\}$$

The set doesn't seem to satisfy either of the two conditions, and therefore it is not a subspace!

I'D IMAGINE YOU MIGHT THINK THAT "BOTH ❶ AND ❷ HOLD IF WE USE $\alpha_1 = \alpha_2 = 0$, SO IT SHOULD BE A SUBSPACE!"

IT'S TRUE THAT THE CONDITIONS HOLD FOR THOSE VALUES, BUT SINCE THE CONDITIONS HAVE TO HOLD FOR ARBITRARY REAL VALUES—THAT IS, *ALL* REAL VALUES—IT'S JUST NOT ENOUGH TO TEST WITH A FEW CHOSEN NUMERICAL EXAMPLES. THE VECTOR SET IS A SUBSPACE ONLY IF BOTH CONDITIONS HOLD FOR ALL KINDS OF VECTORS.

IF THIS STILL DOESN'T MAKE SENSE, DON'T GIVE UP! THIS IS HARD!

THE FOLLOWING SUBSPACES ARE CALLED *LINEAR SPANS* AND ARE A BIT SPECIAL.

WHAT IS A LINEAR SPAN?

We say that a set of m-dimensional vectors

$$\begin{pmatrix} a_{11} \\ a_{21} \\ \vdots \\ a_{m1} \end{pmatrix}, \begin{pmatrix} a_{12} \\ a_{22} \\ \vdots \\ a_{m2} \end{pmatrix}, \ldots, \begin{pmatrix} a_{1n} \\ a_{2n} \\ \vdots \\ a_{mn} \end{pmatrix} \text{ span the following subspace in } R^m:$$

$$\left\{ c_1 \begin{pmatrix} a_{11} \\ a_{21} \\ \vdots \\ a_{m1} \end{pmatrix} + c_2 \begin{pmatrix} a_{12} \\ a_{22} \\ \vdots \\ a_{m2} \end{pmatrix} + \ldots + c_n \begin{pmatrix} a_{1n} \\ a_{2n} \\ \vdots \\ a_{mn} \end{pmatrix} \middle| \begin{array}{l} c_1, c_2, \text{ and } c_n \text{ are} \\ \text{arbitrary numbers} \end{array} \right\}$$

This set forms a subspace and is called the *linear span* of the n original vectors.

EXAMPLE 1

The x_1x_2-plane is a subspace of R^2 and can, for example, be spanned by using the two vectors $\begin{pmatrix} 3 \\ 1 \end{pmatrix}$ and $\begin{pmatrix} 1 \\ 2 \end{pmatrix}$ like so: $\left\{ c_1 \begin{pmatrix} 3 \\ 1 \end{pmatrix} + c_2 \begin{pmatrix} 1 \\ 2 \end{pmatrix} \middle| \begin{array}{l} c_1 \text{ and } c_2 \text{ are} \\ \text{arbitrary numbers} \end{array} \right\}$

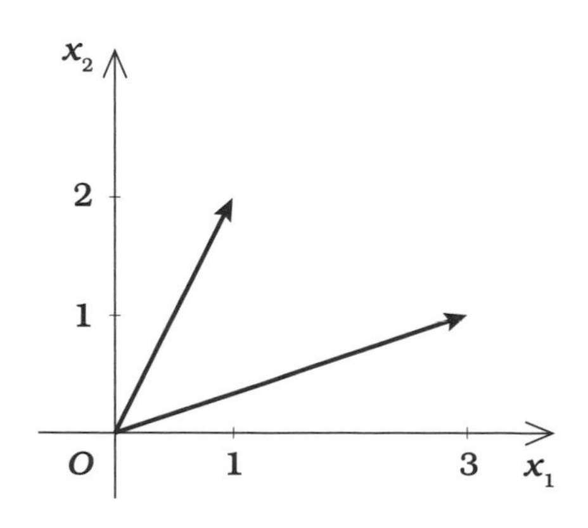

EXAMPLE 2

The x_1x_2-plane could also be a subspace of R^3, and we could span it using the vectors $\begin{pmatrix} 1 \\ 0 \\ 0 \end{pmatrix}$ and $\begin{pmatrix} 0 \\ 1 \\ 0 \end{pmatrix}$, creating this set:

$$\left\{ c_1 \begin{pmatrix} 1 \\ 0 \\ 0 \end{pmatrix} + c_2 \begin{pmatrix} 0 \\ 1 \\ 0 \end{pmatrix} \;\middle|\; \begin{array}{l} c_1 \text{ and } c_2 \text{ are} \\ \text{arbitrary numbers} \end{array} \right\}$$

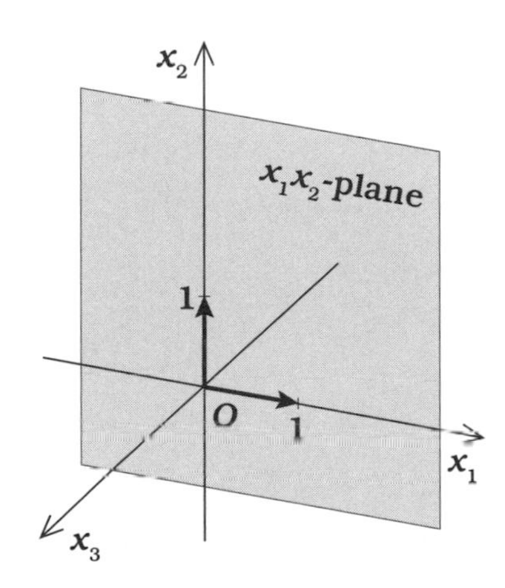

R^m IS ALSO A SUBSPACE OF ITSELF, AS YOU MIGHT HAVE GUESSED FROM EXAMPLE 1.

ALL SUBSPACES CONTAIN THE ZERO FACTOR, WHICH YOU COULD PROBABLY TELL FROM LOOKING AT THE EXAMPLE ON PAGE 152. REMEMBER, THEY MUST PASS THROUGH THE ORIGIN!

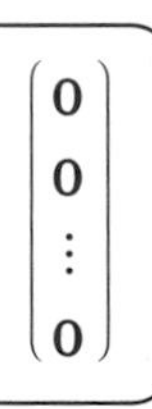

SORRY FOR THE WAIT.

HERE ARE THE DEFINITIONS OF *BASIS* AND *DIMENSION.*

WHAT ARE BASIS AND DIMENSION?

Suppose that W is a subspace of R^m and that it is spanned by the linearly independent vectors $\begin{pmatrix} a_{11} \\ a_{21} \\ \vdots \\ a_{m1} \end{pmatrix}$, $\begin{pmatrix} a_{12} \\ a_{22} \\ \vdots \\ a_{m2} \end{pmatrix}$, and $\begin{pmatrix} a_{1n} \\ a_{2n} \\ \vdots \\ a_{mn} \end{pmatrix}$.

This could also be written as follows:

$$W = \left\{ c_1 \begin{pmatrix} a_{11} \\ a_{21} \\ \vdots \\ a_{m1} \end{pmatrix} + c_2 \begin{pmatrix} a_{12} \\ a_{22} \\ \vdots \\ a_{m2} \end{pmatrix} + \dots + c_n \begin{pmatrix} a_{1n} \\ a_{2n} \\ \vdots \\ a_{mn} \end{pmatrix} \middle| \begin{array}{l} c_1, c_2, \text{ and } c_n \text{ are} \\ \text{arbitrary numbers} \end{array} \right\}$$

When this equality holds, we say that the set $\left\{ \begin{pmatrix} a_{11} \\ a_{21} \\ \vdots \\ a_{m1} \end{pmatrix}, \begin{pmatrix} a_{12} \\ a_{22} \\ \vdots \\ a_{m2} \end{pmatrix}, \dots, \begin{pmatrix} a_{1n} \\ a_{2n} \\ \vdots \\ a_{mn} \end{pmatrix} \right\}$ forms a *basis* to the subspace W.

The *dimension* of the subspace W is equal to the number of vectors in any basis for W.

EXAMPLE

Let's call the x_1x_2-plane W, for simplicity's sake. So suppose that W is a subspace of R^3 and is spanned by the linearly independent vectors

$\begin{pmatrix}3\\1\\0\end{pmatrix}$ and $\begin{pmatrix}1\\2\\0\end{pmatrix}$.

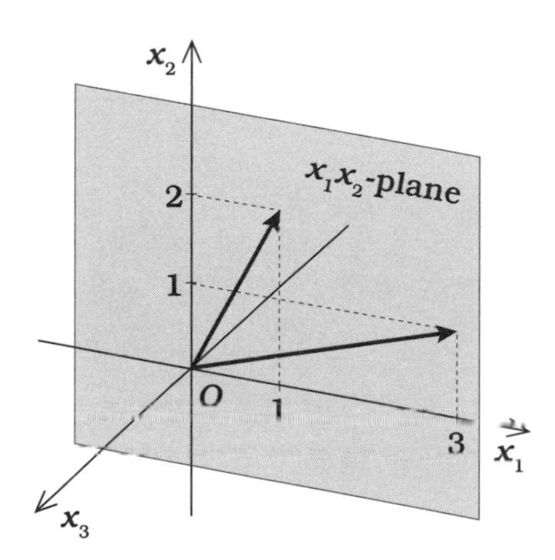

We have this:

$$W = \left\{ c_1\begin{pmatrix}3\\1\\0\end{pmatrix} + c_2\begin{pmatrix}1\\2\\0\end{pmatrix} \middle| \begin{array}{l} c_1 \text{ and } c_2 \text{ are} \\ \text{arbitrary numbers}\end{array} \right\}$$

The fact that this equality holds means that the vector set $\left\{\begin{pmatrix}3\\1\\0\end{pmatrix}, \begin{pmatrix}1\\2\\0\end{pmatrix}\right\}$

is a basis of the subspace W. Since the base contains two vectors, $\dim W = 2$.

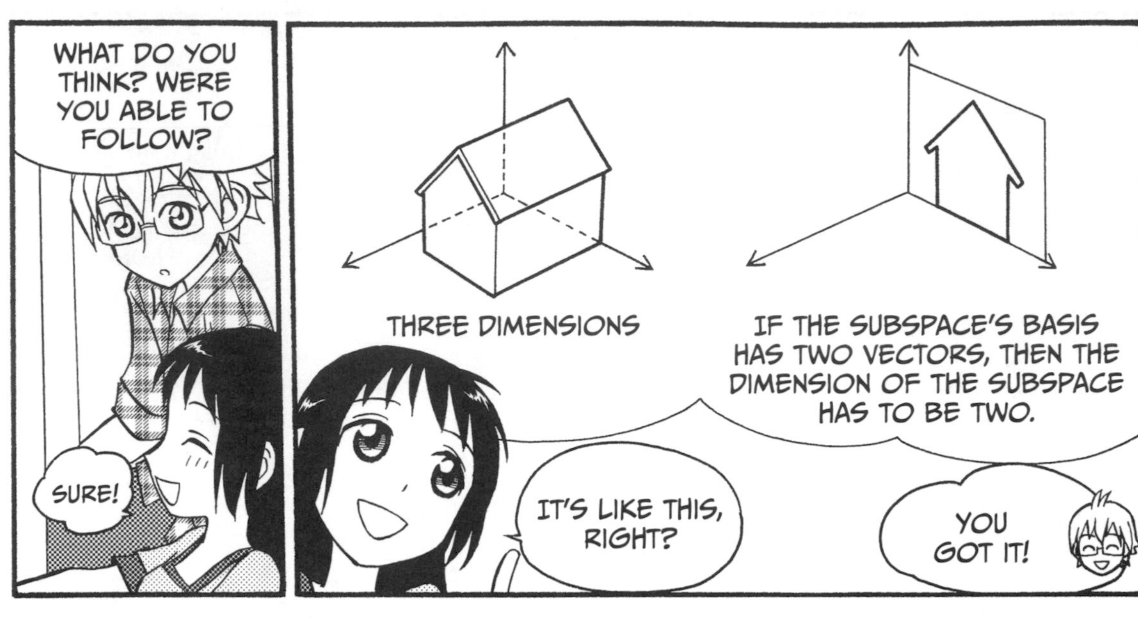
WHAT DO YOU THINK? WERE YOU ABLE TO FOLLOW?
SURE!
THREE DIMENSIONS
IF THE SUBSPACE'S BASIS HAS TWO VECTORS, THEN THE DIMENSION OF THE SUBSPACE HAS TO BE TWO.
IT'S LIKE THIS, RIGHT?
YOU GOT IT!

THAT'S ENOUGH FOR TODAY.
FUN AS ALWAYS!
THANKS FOR ALL THE HELP.

LET'S TALK ABOUT LINEAR TRANSFORMATIONS NEXT TIME.
IT'S ALSO AN IMPORTANT SUBJECT, SO COME PREPARED!
OF COURSE!

OH, LOOKS LIKE WE'RE GOING THE SAME WAY...

...

UMM, I DIDN'T REALLY ANSWER YOUR QUESTION BEFORE...
?

I JOINED THE KARATE CLUB BECAUSE...
I'M TIRED OF BEING SUCH A WIMP.

I WANT TO GET STRONGER.
THAT WAY I CAN...
WELL...

I GUESS YOU'LL BE NEEDING...
...A LOT MORE HOMEMADE LUNCHES, THEN!

WHA-?

TO SURVIVE MY BROTHER'S REIGN OF TERROR, THAT IS. DON'T WORRY, I'LL MAKE YOU MY SUPER SPECIAL STAMINA-LUNCH EXTRAVAGANZA EVERY WEEK FROM NOW ON!

THANK YOU, MISA.
HEHE, DON'T WORRY 'BOUT IT!

COORDINATES

Coordinates in linear algebra are a bit different from the coordinates explained in high school. I'll try explaining the difference between the two using the image below.

When working with coordinates and coordinate systems at the high school level, it's much easier to use only the trivial basis:

$$\left\{\begin{pmatrix}1\\0\\\vdots\\0\end{pmatrix}, \begin{pmatrix}0\\1\\\vdots\\0\end{pmatrix}, \dots, \begin{pmatrix}0\\0\\\vdots\\1\end{pmatrix}\right\}$$

In this kind of system, the relationship between the origin and the point in the top right is interpreted as follows:

$$\begin{pmatrix}7\\4\end{pmatrix} = 7\begin{pmatrix}1\\0\end{pmatrix} + 4\begin{pmatrix}0\\1\end{pmatrix}$$

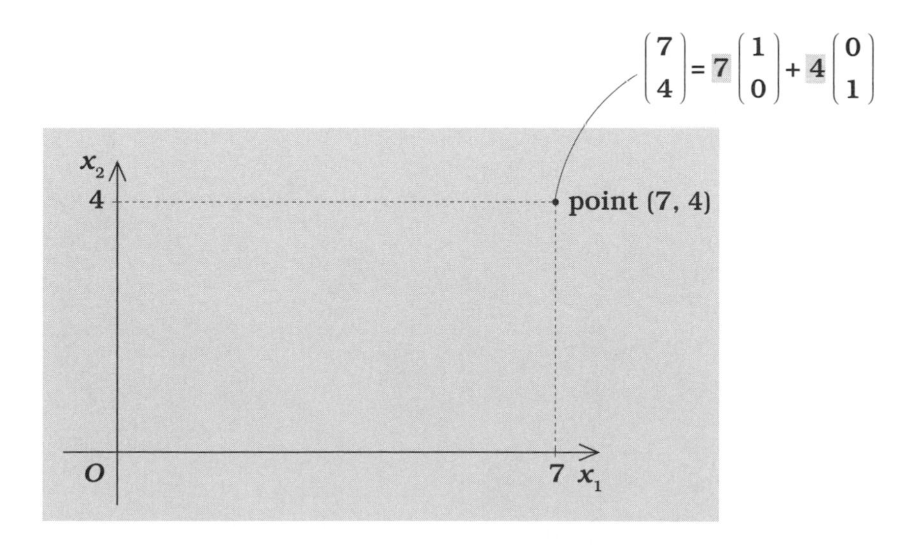

It is important to understand that the trivial basis is only one of many bases when we move into the realm of linear algebra—and that using other bases produces other relationships between the origin and a given point. The image below illustrates the point (2, 1) in a system using the nontrivial basis consisting of the two vectors $u_1 = \begin{pmatrix} 3 \\ 1 \end{pmatrix}$ and $u_2 = \begin{pmatrix} 1 \\ 2 \end{pmatrix}$.

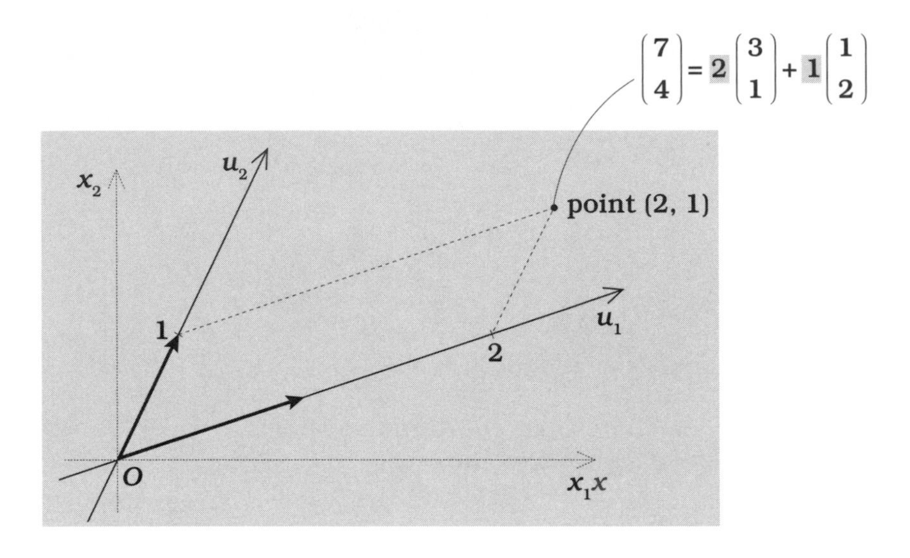

This alternative way of thinking about coordinates is very useful in factor analysis, for example.

7
LINEAR TRANSFORMATIONS

A PRACTICE MATCH WITH NANHOU UNIVERSITY?
THAT'S RIGHT! WE GO HEAD-TO-HEAD IN TWO WEEKS.

A MATCH, HUH? I GUESS I'LL BE SITTING IT OUT.
YURINO!
YOU'RE IN.
WHAT?!

FOR REAL?
UM, SENSEI? ISN'T IT A BIT EARLY FOR...
ARE YOU TELLING ME WHAT TO DO?
OH NO! OF COURSE NOT!
SLAM

I'LL BE LOOKING FORWARD TO SEEING WHAT YOU'VE LEARNED SO FAR!
UNDERSTOOD?
OSSU!
OF COURSE!
GREAT. DISMISSED!
A MATCH...
WHAT AM I GOING TO DO...
SHUDDER
STOP THINKING LIKE THAT! HE'S GIVING ME THIS OPPORTUNITY.
I HAVE TO DO MY BEST.

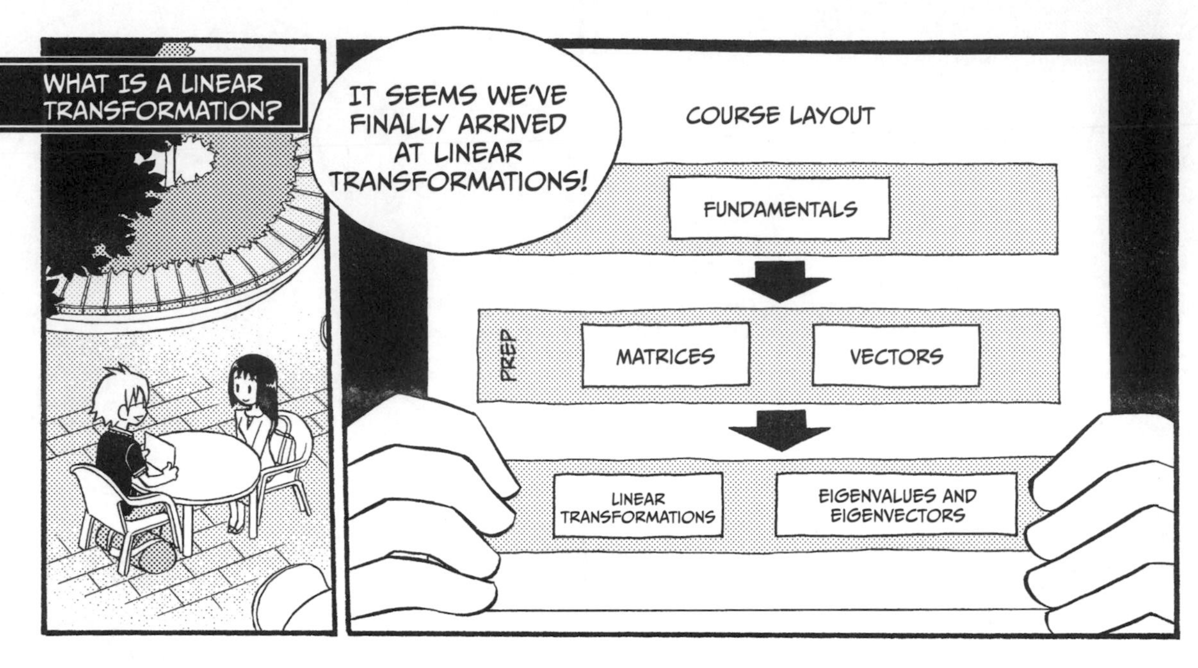

LINEAR TRANSFORMATIONS

Let x_i and x_j be two arbitrary elements, c an arbitrary real number, and f a function from X to Y.

We say that f is a linear transformation from X to Y if it satisfies the following two conditions:

❶ **$f(x_i) + f(x_j)$ and $f(x_i + x_j)$ are equal**

❷ **$cf(x_i)$ and $f(cx_i)$ are equal**

BUT THIS DEFINITION IS ACTUALLY INCOMPLETE.

LINEAR TRANSFORMATIONS

Let $\begin{pmatrix} x_{1i} \\ x_{2i} \\ \vdots \\ x_{ni} \end{pmatrix}$ and $\begin{pmatrix} x_{1j} \\ x_{2j} \\ \vdots \\ x_{nj} \end{pmatrix}$ be two arbitrary elements from R^n, c an arbitrary real number, and f a function from R^n to R^m.

We say that f is a linear transformation from R^n to R^m if it satisfies the following two conditions:

❶ $f\left(\begin{pmatrix} x_{1i} \\ x_{2i} \\ \vdots \\ x_{ni} \end{pmatrix}\right) + f\left(\begin{pmatrix} x_{1j} \\ x_{2j} \\ \vdots \\ x_{nj} \end{pmatrix}\right)$ and $f\left(\begin{pmatrix} x_{1i} + x_{1j} \\ x_{2i} + x_{2j} \\ \vdots \\ x_{ni} + x_{nj} \end{pmatrix}\right)$ are equal.

❷ $cf\left(\begin{pmatrix} x_{1i} \\ x_{2i} \\ \vdots \\ x_{ni} \end{pmatrix}\right)$ and $f\left(c\begin{pmatrix} x_{1i} \\ x_{2i} \\ \vdots \\ x_{ni} \end{pmatrix}\right)$ are equal.

A linear transformation from R^n to R^m is sometimes called a *linear map* or *linear operation*.

AND IF f IS A LINEAR TRANSFORMATION FROM R^n TO R^m...
$\mathbb{R}^n$
f
$\mathbb{R}^m$

THEN IT SHOULDN'T BE A SURPRISE TO HEAR THAT f CAN BE WRITTEN AS AN $m \times n$ MATRIX.
UM...IT SHOULDN'T?
HAVE A LOOK AT THE FOLLOWING EQUATIONS.
$\mathbb{R}^2$
$\begin{pmatrix} a_{11} & a_{12} \\ a_{21} & a_{22} \\ a_{31} & a_{32} \end{pmatrix}$
$\mathbb{R}^3$
$\mathbb{R}^3$
$\begin{pmatrix} a_{11} & a_{12} & a_{13} \\ a_{21} & a_{22} & a_{23} \end{pmatrix}$
$\mathbb{R}^2$
$\mathbb{R}^2$
$\begin{pmatrix} a_{11} & a_{12} \\ a_{21} & a_{22} \end{pmatrix}$
$\mathbb{R}^2$

❶ We'll verify the first rule first: $$f\left(\begin{pmatrix} x_{1i} \\ x_{2i} \\ \vdots \\ x_{ni} \end{pmatrix}\right) + f\left(\begin{pmatrix} x_{1j} \\ x_{2j} \\ \vdots \\ x_{nj} \end{pmatrix}\right) = f\left(\begin{pmatrix} x_{1i} + x_{1j} \\ x_{2i} + x_{2j} \\ \vdots \\ x_{ni} + x_{nj} \end{pmatrix}\right)$$

We just replace f with a matrix, then simplify:

$$\begin{pmatrix} a_{11} & a_{12} & \cdots & a_{1n} \\ a_{21} & a_{22} & \cdots & a_{2n} \\ \vdots & \vdots & \ddots & \vdots \\ a_{m1} & a_{m2} & \cdots & a_{mn} \end{pmatrix} \begin{pmatrix} x_{1i} \\ x_{2i} \\ \vdots \\ x_{ni} \end{pmatrix} + \begin{pmatrix} a_{11} & a_{12} & \cdots & a_{1n} \\ a_{21} & a_{22} & \cdots & a_{2n} \\ \vdots & \vdots & \ddots & \vdots \\ a_{m1} & a_{m2} & \cdots & a_{mn} \end{pmatrix} \begin{pmatrix} x_{1j} \\ x_{2j} \\ \vdots \\ x_{nj} \end{pmatrix}$$

$$= \begin{pmatrix} a_{11}x_{1i} + a_{12}x_{2i} + \dots + a_{1n}x_{ni} \\ a_{21}x_{1i} + a_{22}x_{2i} + \dots + a_{2n}x_{ni} \\ \vdots \\ a_{m1}x_{1i} + a_{m2}x_{2i} + \dots + a_{mn}x_{ni} \end{pmatrix} + \begin{pmatrix} a_{11}x_{1j} + a_{12}x_{2j} + \dots + a_{1n}x_{nj} \\ a_{21}x_{1j} + a_{22}x_{2j} + \dots + a_{2n}x_{nj} \\ \vdots \\ a_{m1}x_{1j} + a_{m2}x_{2j} + \dots + a_{mn}x_{nj} \end{pmatrix}$$

$$= \begin{pmatrix} a_{11}(x_{1i} + x_{1j}) + a_{12}(x_{2i} + x_{2j}) + \dots + a_{1n}(x_{ni} + x_{nj}) \\ a_{21}(x_{1i} + x_{1j}) + a_{22}(x_{2i} + x_{2j}) + \dots + a_{2n}(x_{ni} + x_{nj}) \\ \vdots \\ a_{m1}(x_{1i} + x_{1j}) + a_{m2}(x_{2i} + x_{2j}) + \dots + a_{mn}(x_{ni} + x_{nj}) \end{pmatrix}$$

$$= \begin{pmatrix} a_{11} & a_{12} & \cdots & a_{1n} \\ a_{21} & a_{22} & \cdots & a_{2n} \\ \vdots & \vdots & \ddots & \vdots \\ a_{m1} & a_{m2} & \cdots & a_{mn} \end{pmatrix} \begin{pmatrix} x_{1i} + x_{1j} \\ x_{2i} + x_{2j} \\ \vdots \\ x_{ni} + x_{nj} \end{pmatrix}$$

❷ Now for the second rule: $$cf\left(\begin{pmatrix} x_{1i} \\ x_{2i} \\ \vdots \\ x_{ni} \end{pmatrix}\right) = f\left(c\begin{pmatrix} x_{1i} \\ x_{2i} \\ \vdots \\ x_{ni} \end{pmatrix}\right)$$

Again, just replace f with a matrix and simplify:

$$c\begin{pmatrix} a_{11} & a_{12} & \cdots & a_{1n} \\ a_{21} & a_{22} & \cdots & a_{2n} \\ \vdots & \vdots & \ddots & \vdots \\ a_{m1} & a_{m2} & \cdots & a_{mn} \end{pmatrix}\begin{pmatrix} x_{1i} \\ x_{2i} \\ \vdots \\ x_{ni} \end{pmatrix}$$

$$= c\begin{pmatrix} a_{11}x_{1i} + a_{12}x_{2i} + \ldots + a_{1n}x_{ni} \\ a_{21}x_{1i} + a_{22}x_{2i} + \ldots + a_{2n}x_{ni} \\ \vdots \\ a_{m1}x_{1i} + a_{m2}x_{2i} + \ldots + a_{mn}x_{ni} \end{pmatrix}$$

$$= \begin{pmatrix} a_{11}(cx_{1i}) + a_{12}(cx_{2i}) + \ldots + a_{1n}(cx_{ni}) \\ a_{21}(cx_{1i}) + a_{22}(cx_{2i}) + \ldots + a_{2n}(cx_{ni}) \\ \vdots \\ a_{m1}(cx_{1i}) + a_{m2}(cx_{2i}) + \ldots + a_{mn}(cx_{ni}) \end{pmatrix}$$

$$= \begin{pmatrix} a_{11} & a_{12} & \cdots & a_{1n} \\ a_{21} & a_{22} & \cdots & a_{2n} \\ \vdots & \vdots & \ddots & \vdots \\ a_{m1} & a_{m2} & \cdots & a_{mn} \end{pmatrix}\begin{pmatrix} cx_{1i} \\ cx_{2i} \\ \vdots \\ cx_{ni} \end{pmatrix}$$

$$= \begin{pmatrix} a_{11} & a_{12} & \cdots & a_{1n} \\ a_{21} & a_{22} & \cdots & a_{2n} \\ \vdots & \vdots & \ddots & \vdots \\ a_{m1} & a_{m2} & \cdots & a_{mn} \end{pmatrix}\left[c\begin{pmatrix} x_{1i} \\ x_{2i} \\ \vdots \\ x_{ni} \end{pmatrix}\right]$$

❶ **We'll show that the first rule holds:**

$$\begin{pmatrix} a_{11} & a_{12} \\ a_{21} & a_{22} \end{pmatrix}\begin{pmatrix} x_{1i} \\ x_{2i} \end{pmatrix} + \begin{pmatrix} a_{11} & a_{12} \\ a_{21} & a_{22} \end{pmatrix}\begin{pmatrix} x_{1j} \\ x_{2j} \end{pmatrix} = \begin{pmatrix} a_{11} & a_{12} \\ a_{21} & a_{22} \end{pmatrix}\begin{pmatrix} x_{1i} + x_{1j} \\ x_{2i} + x_{2j} \end{pmatrix}$$

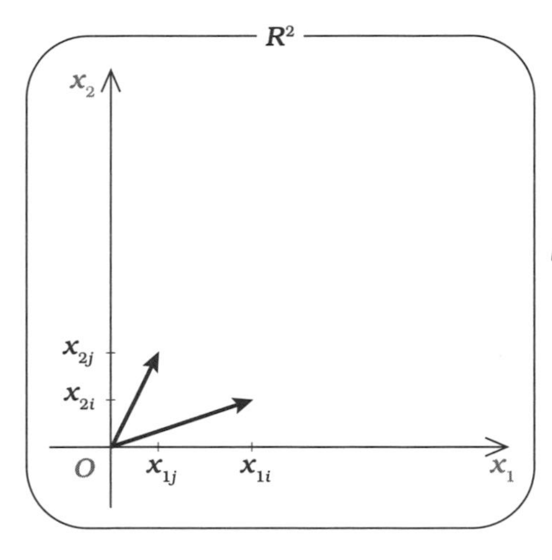

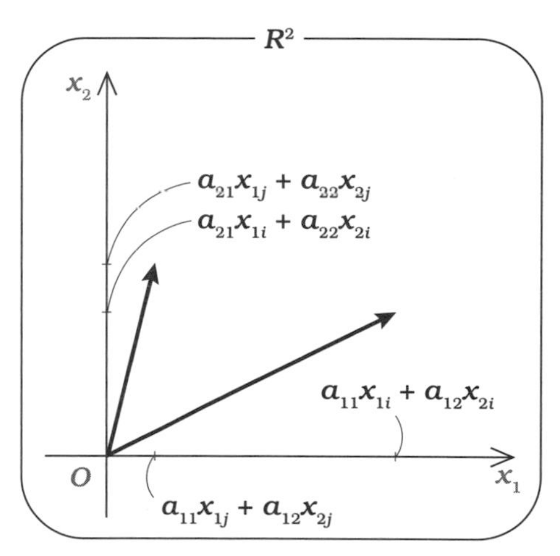

IF YOU ADD FIRST...

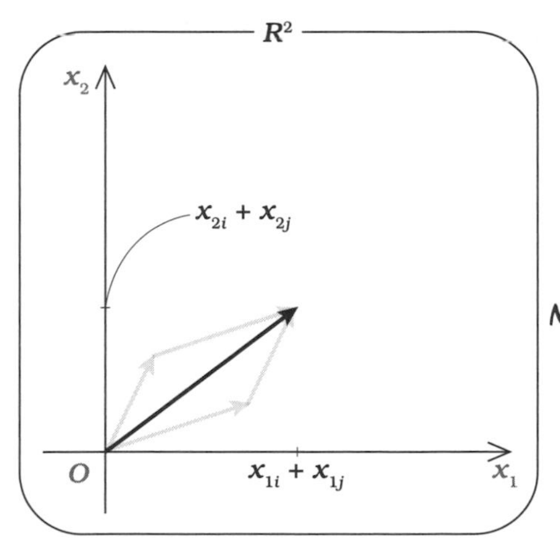

THEN MULTIPLY...

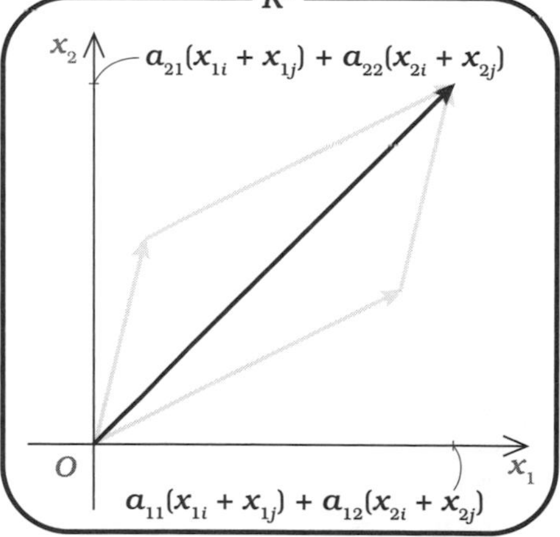

YOU GET THE SAME FINAL RESULT!

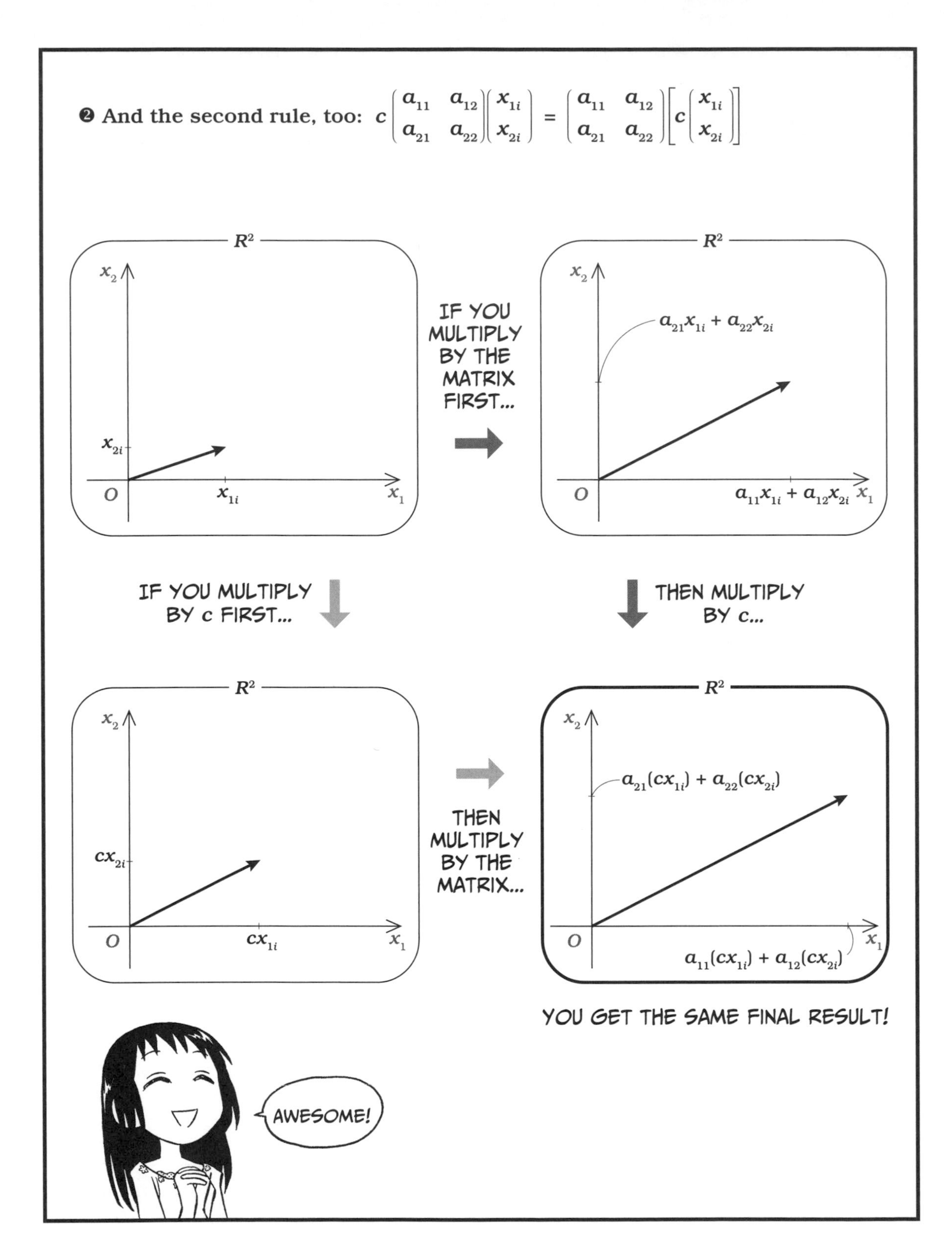

❷ And the second rule, too: $c\begin{pmatrix} a_{11} & a_{12} \\ a_{21} & a_{22} \end{pmatrix}\begin{pmatrix} x_{1i} \\ x_{2i} \end{pmatrix} = \begin{pmatrix} a_{11} & a_{12} \\ a_{21} & a_{22} \end{pmatrix}\left[c\begin{pmatrix} x_{1i} \\ x_{2i} \end{pmatrix}\right]$
R^2
x_2
x_{2i}
O
x_{1i}
x_1
IF YOU MULTIPLY BY THE MATRIX FIRST...
R^2
x_2
$a_{21}x_{1i} + a_{22}x_{2i}$
O
$a_{11}x_{1i} + a_{12}x_{2i}$
x_1
IF YOU MULTIPLY BY c FIRST...
THEN MULTIPLY BY c...
R^2
x_2
cx_{2i}
O
cx_{1i}
x_1
THEN MULTIPLY BY THE MATRIX...
R^2
x_2
$a_{21}(cx_{1i}) + a_{22}(cx_{2i})$
O
x_1
$a_{11}(cx_{1i}) + a_{12}(cx_{2i})$
YOU GET THE SAME FINAL RESULT!
AWESOME!

So when f is a linear transformation from R^n to R^m, we can also say that f is equivalent to the $m \times n$ matrix that defines the linear transformation from R^n to R^m.

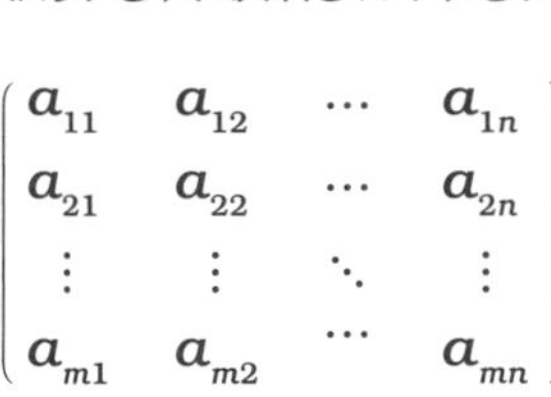

$$\begin{pmatrix} a_{11} & a_{12} & \cdots & a_{1n} \\ a_{21} & a_{22} & \cdots & a_{2n} \\ \vdots & \vdots & \ddots & \vdots \\ a_{m1} & a_{m2} & \cdots & a_{mn} \end{pmatrix}$$

NOW I GET IT!

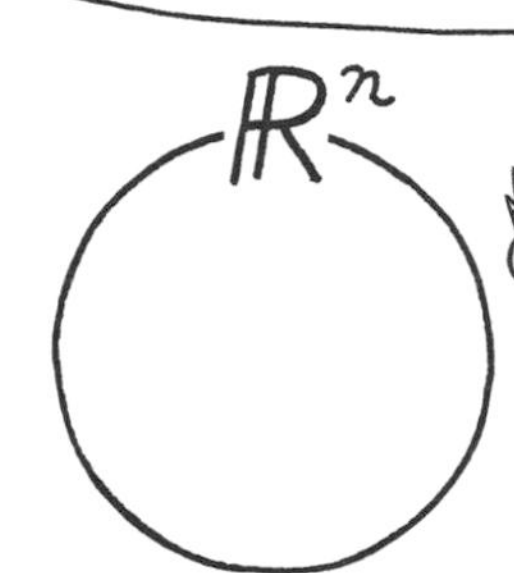

WHY WE STUDY LINEAR TRANSFORMATIONS

SO...WHAT ARE LINEAR TRANSFORMATIONS GOOD FOR, EXACTLY?

THEY SEEM PRETTY IMPORTANT. I GUESS WE'LL BE USING THEM A LOT FROM NOW ON?

WELL, IT'S NOT REALLY A QUESTION OF IMPORTANCE...

SO WHY DO WE HAVE TO STUDY THEM?

WELL...

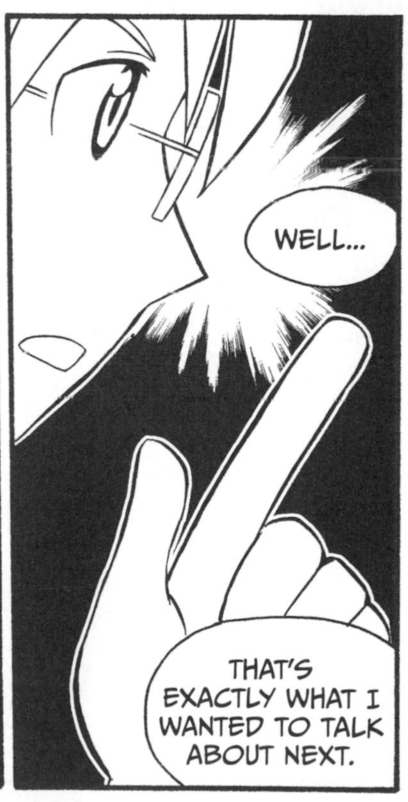

THAT'S EXACTLY WHAT I WANTED TO TALK ABOUT NEXT.

CONSIDER THE LINEAR TRANSFORMATION FROM R^n TO R^m DEFINED BY THE FOLLOWING $m \times n$ MATRIX:

$$\begin{pmatrix} a_{11} & a_{12} & \cdots & a_{1n} \\ a_{21} & a_{22} & \cdots & a_{2n} \\ \vdots & \vdots & \ddots & \vdots \\ a_{m1} & a_{m2} & \cdots & a_{mn} \end{pmatrix}$$

IF $\begin{pmatrix} y_1 \\ y_2 \\ \vdots \\ y_m \end{pmatrix}$ IS THE IMAGE OF $\begin{pmatrix} x_1 \\ x_2 \\ \vdots \\ x_n \end{pmatrix}$ UNDER THIS LINEAR TRANSFORMATION,

THEN THE FOLLOWING EQUATION IS TRUE:

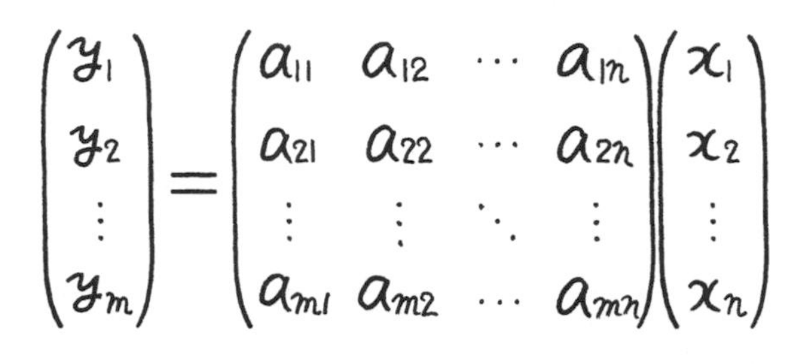

$$\begin{pmatrix} y_1 \\ y_2 \\ \vdots \\ y_m \end{pmatrix} = \begin{pmatrix} a_{11} & a_{12} & \cdots & a_{1n} \\ a_{21} & a_{22} & \cdots & a_{2n} \\ \vdots & \vdots & \ddots & \vdots \\ a_{m1} & a_{m2} & \cdots & a_{mn} \end{pmatrix} \begin{pmatrix} x_1 \\ x_2 \\ \vdots \\ x_n \end{pmatrix}$$

IMAGES

Suppose x_i is an element from X.

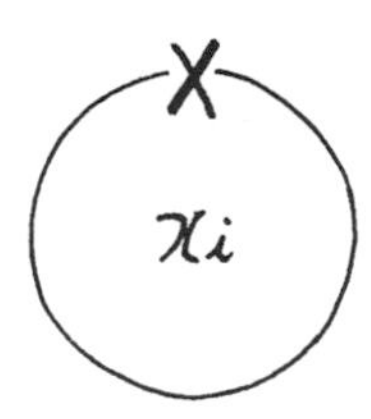

WE TALKED A BIT ABOUT THIS BEFORE, DIDN'T WE?

The element in Y corresponding to x_i under f is called "x_i's image under f."

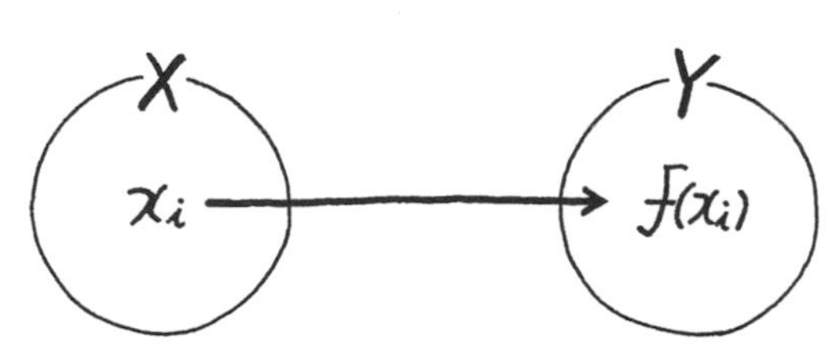

YEAH, IN CHAPTER 2.

$$\begin{pmatrix} y_1 \\ y_2 \\ \vdots \\ y_m \end{pmatrix} = \begin{pmatrix} a_{11} & a_{12} & \cdots & a_{1n} \\ a_{21} & a_{22} & \cdots & a_{2n} \\ \vdots & \vdots & \ddots & \vdots \\ a_{m1} & a_{m2} & \cdots & a_{mn} \end{pmatrix} \begin{pmatrix} x_1 \\ x_2 \\ \vdots \\ x_n \end{pmatrix}$$

$$y = a x$$

$$\begin{pmatrix} x_1 \\ x_2 \\ \vdots \\ x_n \end{pmatrix} \downarrow \begin{pmatrix} a_{11} & a_{12} & \cdots & a_{1n} \\ a_{21} & a_{22} & \cdots & a_{2n} \\ \vdots & \vdots & \ddots & \vdots \\ a_{m1} & a_{m2} & \cdots & a_{mn} \end{pmatrix} \begin{pmatrix} x_1 \\ x_2 \\ \vdots \\ x_n \end{pmatrix} \downarrow \begin{pmatrix} y_1 \\ y_2 \\ \vdots \\ y_m \end{pmatrix}$$

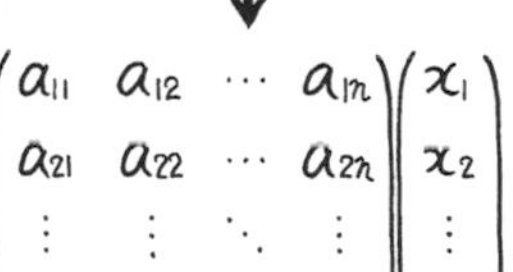

Multiplying an *n*-dimensional space by an *m*×*n* matrix...

$$\begin{pmatrix} a_{11} & a_{12} & \cdots & a_{1n} \\ a_{21} & a_{22} & \cdots & a_{2n} \\ \vdots & \vdots & \ddots & \vdots \\ a_{m1} & a_{m2} & \cdots & a_{mn} \end{pmatrix}$$

turns it *m*-dimensional!

I HAVE TO LEARN THIS STUFF BECAUSE OF...THAT?

OOH, BUT "THAT" IS A LOT MORE SIGNIFICANT THAN YOU MIGHT THINK!

TAKE THIS LINEAR TRANSFORMATION FROM THREE TO TWO DIMENSIONS, FOR EXAMPLE.

$$\begin{pmatrix} y_1 \\ y_2 \end{pmatrix} = \begin{pmatrix} a_{11} & a_{12} & a_{13} \\ a_{21} & a_{22} & a_{23} \end{pmatrix} \begin{pmatrix} x_1 \\ x_2 \\ x_3 \end{pmatrix}$$

YOU COULD WRITE IT AS THIS LINEAR SYSTEM OF EQUATIONS INSTEAD, IF YOU WANTED TO.

$$\begin{cases} y_1 = a_{11}x_1 + a_{12}x_2 + a_{13}x_3 \\ y_2 = a_{21}x_1 + a_{22}x_2 + a_{23}x_3 \end{cases}$$

BUT YOU HAVE TO AGREE THAT THIS DOESN'T REALLY CONVEY THE FEELING OF "TRANSFORMING A THREE-DIMENSIONAL SPACE INTO A TWO-DIMENSIONAL ONE," RIGHT?

$\begin{cases} y_1 = a_{11}x_1 + a_{12}x_2 + a_{13}x_3 \\ y_2 = a_{21}x_1 + a_{22}x_2 + a_{23}x_3 \end{cases}$
IS THE SAME AS...
?
I WONDER WHAT THESE LINEAR SYSTEMS ARE SUPPOSED TO BE?
AH...

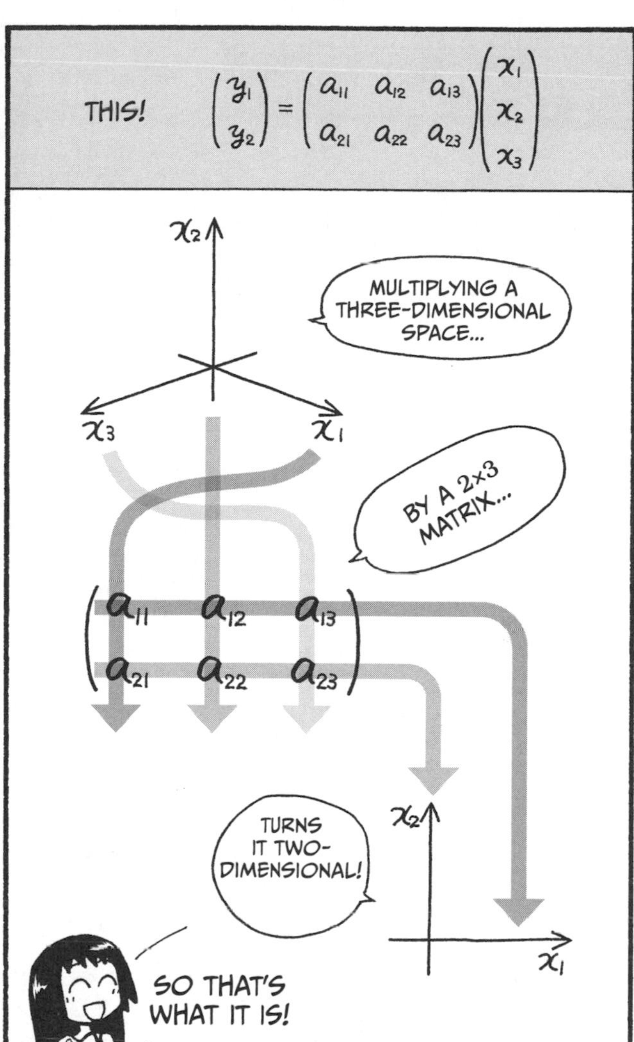

THIS!
$\begin{pmatrix} y_1 \\ y_2 \end{pmatrix} = \begin{pmatrix} a_{11} & a_{12} & a_{13} \\ a_{21} & a_{22} & a_{23} \end{pmatrix} \begin{pmatrix} x_1 \\ x_2 \\ x_3 \end{pmatrix}$
x_2
MULTIPLYING A THREE-DIMENSIONAL SPACE...
x_3
x_1
BY A 2×3 MATRIX...
$\begin{pmatrix} a_{11} & a_{12} & a_{13} \\ a_{21} & a_{22} & a_{23} \end{pmatrix}$
x_2
TURNS IT TWO-DIMENSIONAL!
x_1
SO THAT'S WHAT IT IS!

I THINK I'M STARTING TO GET IT.
LINEAR TRANSFORMATIONS ARE DEFINITELY ONE OF THE HARDER-TO-UNDERSTAND PARTS OF LINEAR ALGEBRA. I REMEMBER HAVING TROUBLE WITH THEM WHEN I STARTED STUDYING THE SUBJECT.

SPECIAL TRANSFORMATIONS
I WOULDN'T WANT YOU THINKING THAT LINEAR TRANSFORMATIONS LACK PRACTICAL USES, THOUGH. COMPUTER GRAPHICS, FOR EXAMPLE, RELY HEAVILY ON LINEAR ALGEBRA AND LINEAR TRANSFORMATIONS IN PARTICULAR.
REALLY?

YEAH. AS WE'RE ALREADY ON THE SUBJECT, LET'S HAVE A LOOK AT SOME OF THE TRANSFORMATIONS THAT LET US DO THINGS LIKE SCALING, ROTATION, TRANSLATION, AND 3-D PROJECTION.
CLICK
AWW! CUTE!
LET'S USE ONE OF MY DRAWINGS.

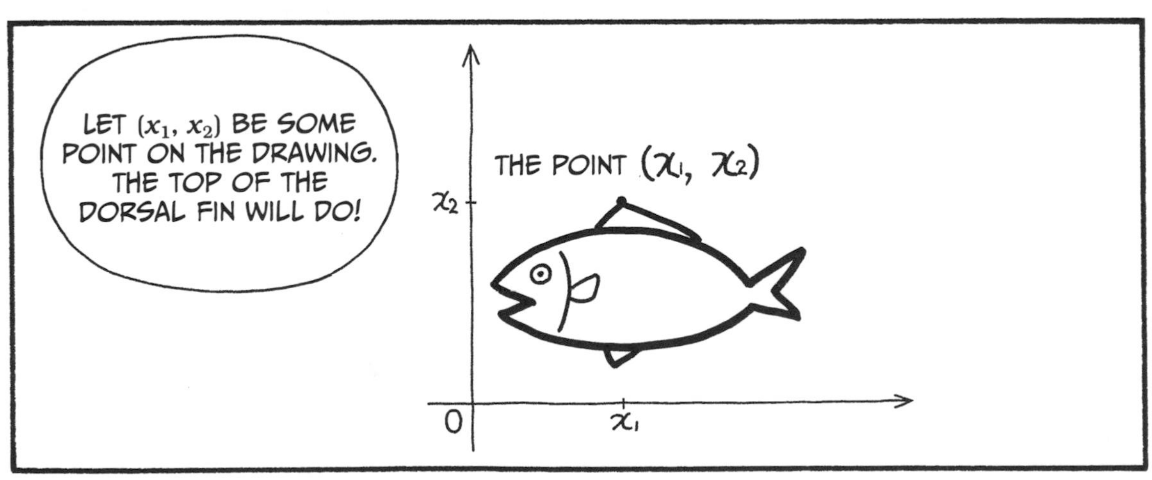
LET (x_1, x_2) BE SOME POINT ON THE DRAWING. THE TOP OF THE DORSAL FIN WILL DO!
THE POINT (x_1, x_2)
x_2
0
x_1

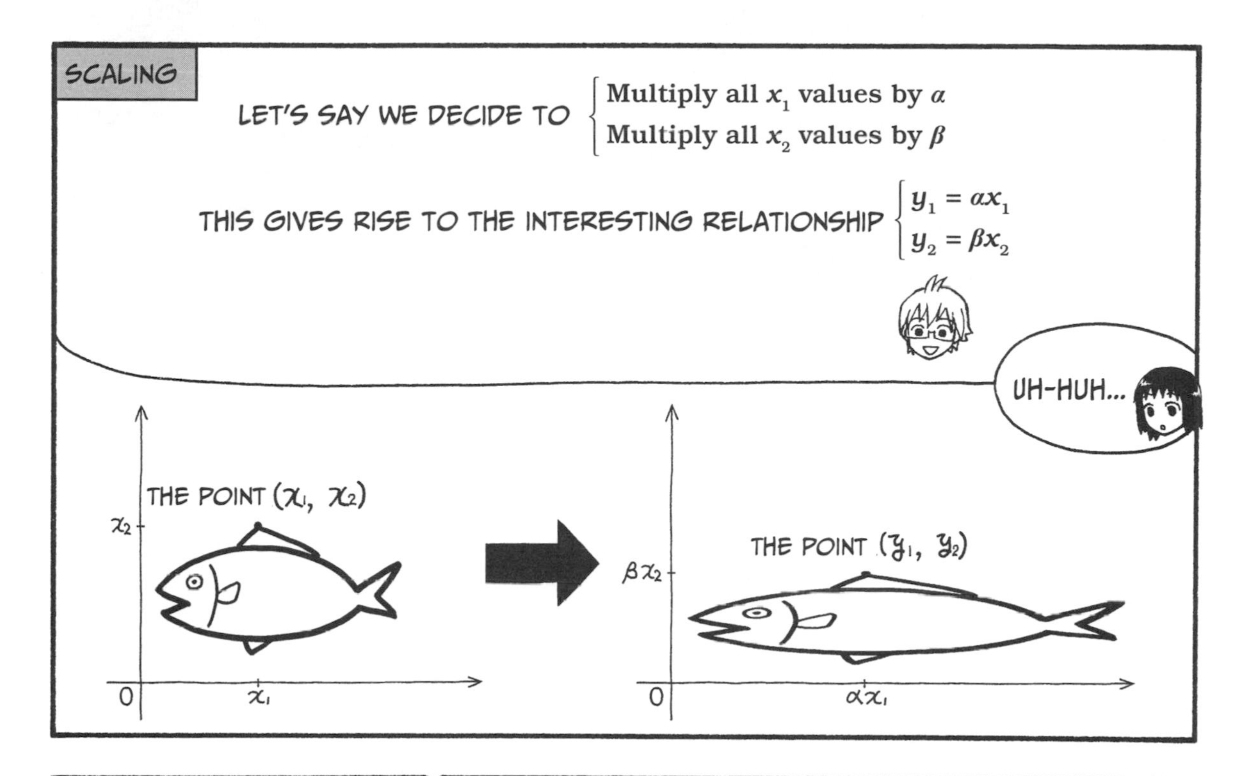

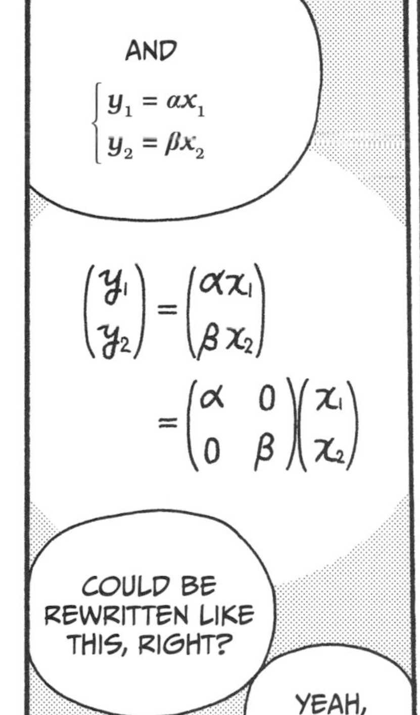

SO THAT MEANS THAT APPLYING THE SET OF RULES

Multiply all x_1 values by α
Multiply all x_2 values by β

ONTO AN ARBITRARY IMAGE IS BASICALLY THE SAME THING AS PASSING THE IMAGE THROUGH A LINEAR TRANSFORMATION IN R^2 EQUAL TO THE FOLLOWING MATRIX!

$$\begin{pmatrix} \alpha & 0 \\ 0 & \beta \end{pmatrix}$$

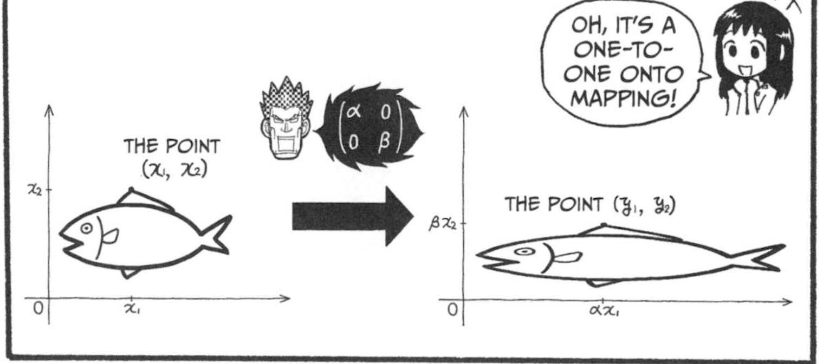

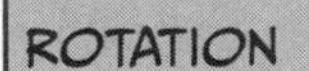

- **Rotating** $\begin{pmatrix} x_1 \\ 0 \end{pmatrix}$ **by** θ* **degrees gets us** $\begin{pmatrix} x_1\cos\theta \\ x_1\sin\theta \end{pmatrix}$

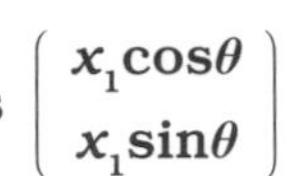

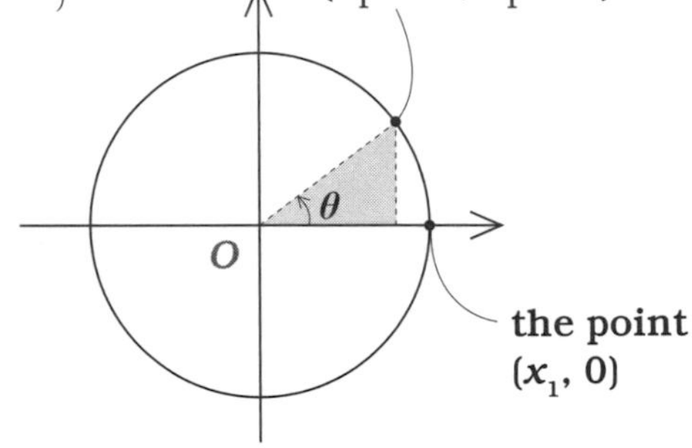

- **Rotating** $\begin{pmatrix} 0 \\ x_2 \end{pmatrix}$ **by** θ **degrees gets us** $\begin{pmatrix} -x_2\sin\theta \\ x_2\cos\theta \end{pmatrix}$

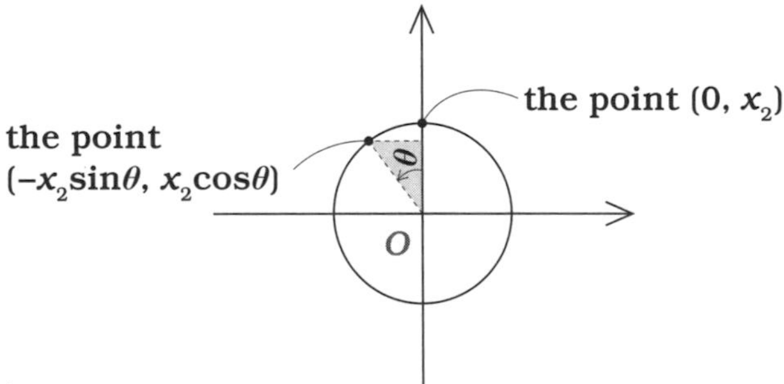

- **Rotating** $\begin{pmatrix} x_1 \\ x_2 \end{pmatrix}$, **that is** $\begin{pmatrix} x_1 \\ 0 \end{pmatrix} + \begin{pmatrix} 0 \\ x_2 \end{pmatrix}$,

  by θ **degrees gets us**

$$\begin{pmatrix} x_1\cos\theta \\ x_1\sin\theta \end{pmatrix} + \begin{pmatrix} -x_2\sin\theta \\ x_2\cos\theta \end{pmatrix}$$

$$= \begin{pmatrix} x_1\cos\theta - x_2\sin\theta \\ x_1\sin\theta + x_2\cos\theta \end{pmatrix}$$

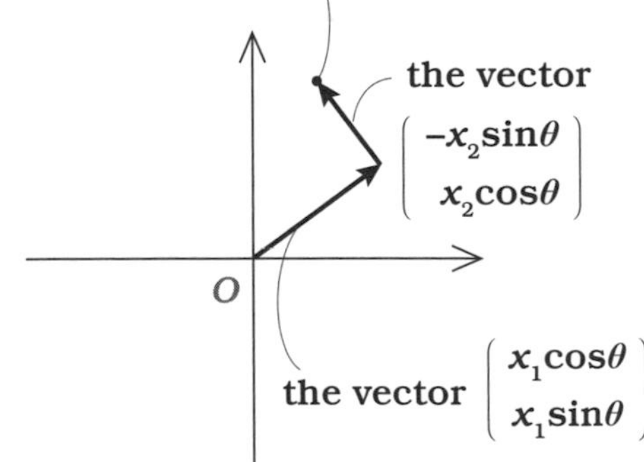

* θ is the Greek letter *theta*.

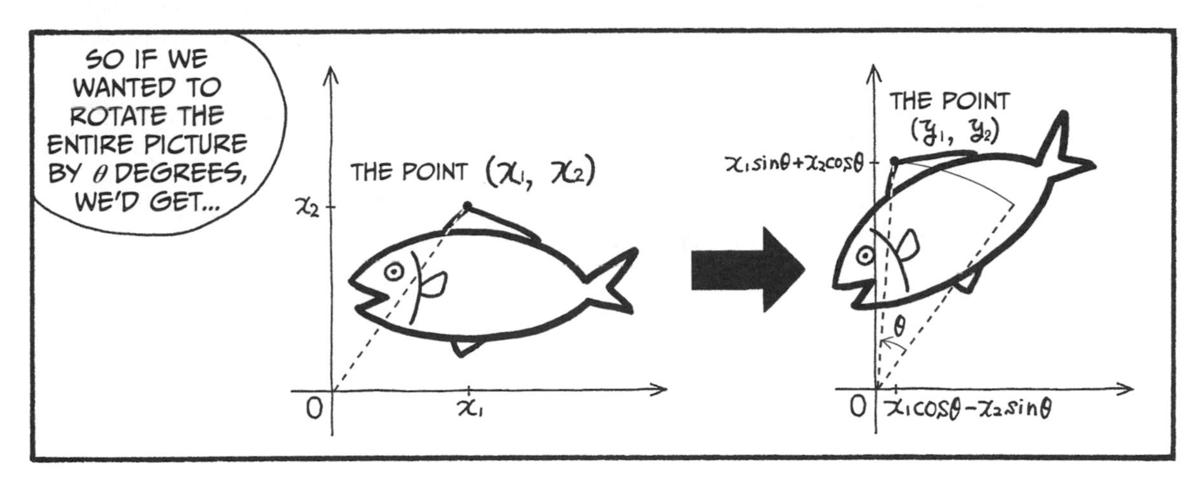

...DUE TO THIS RELATIONSHIP.

$$\begin{pmatrix} y_1 \\ y_2 \end{pmatrix} = \begin{pmatrix} x_1\cos\theta - x_2\sin\theta \\ x_1\sin\theta + x_2\cos\theta \end{pmatrix} = \begin{pmatrix} \cos\theta & -\sin\theta \\ \sin\theta & \cos\theta \end{pmatrix} \begin{pmatrix} x_1 \\ x_2 \end{pmatrix}$$

AHA.

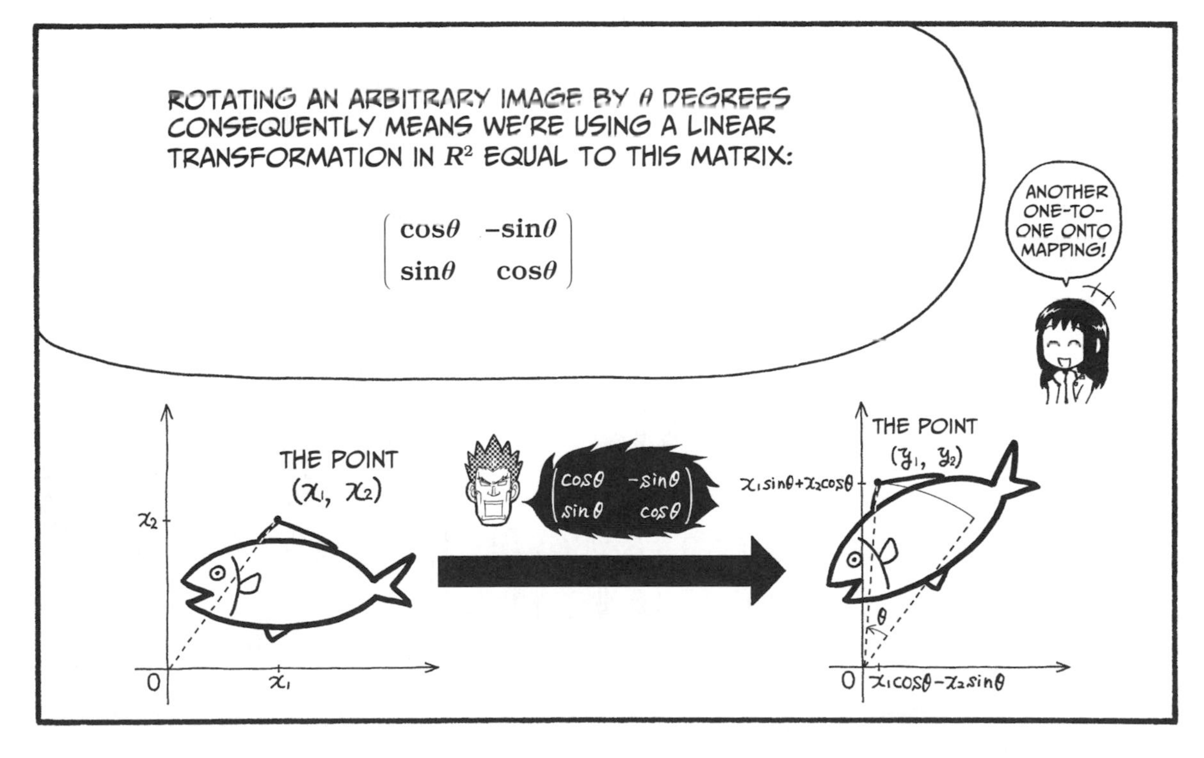

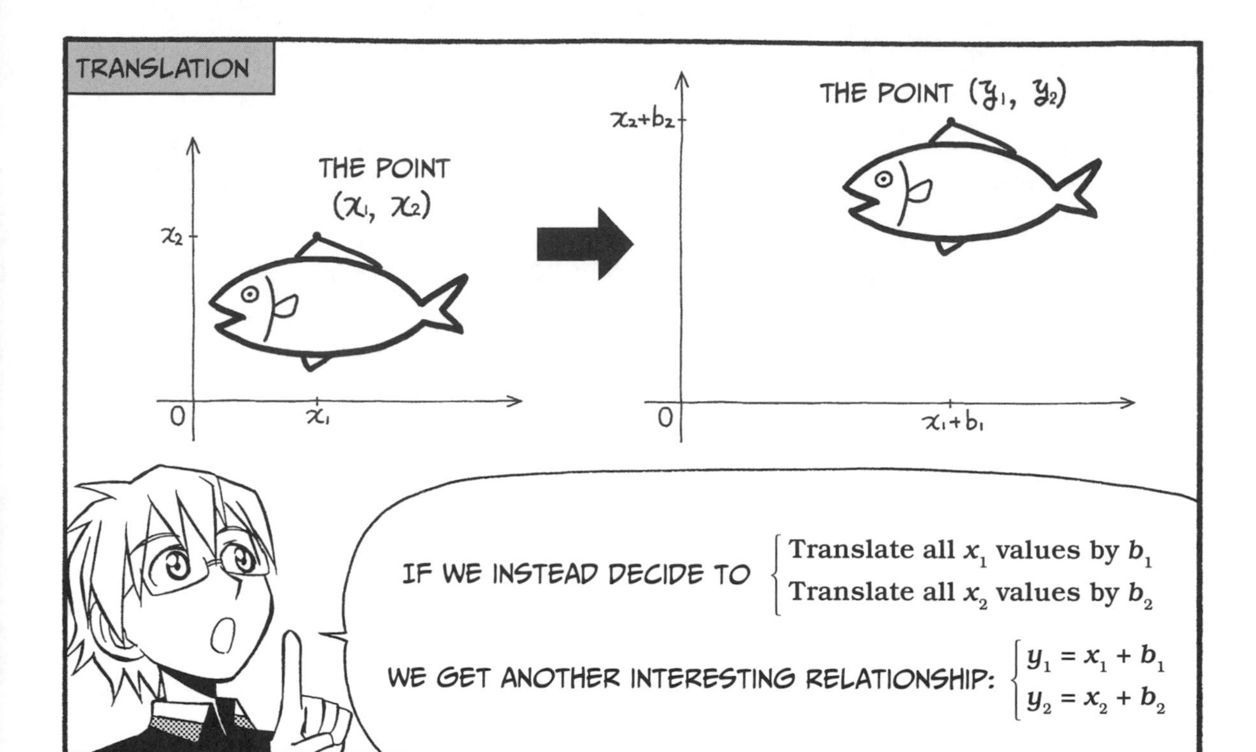

AND THIS CAN ALSO BE REWRITTEN LIKE SO:

$$\begin{pmatrix} y_1 \\ y_2 \end{pmatrix} = \begin{pmatrix} x_1 + b_1 \\ x_2 + b_2 \end{pmatrix} = \begin{pmatrix} 1 & 0 \\ 0 & 1 \end{pmatrix}\begin{pmatrix} x_1 \\ x_2 \end{pmatrix} + \begin{pmatrix} b_1 \\ b_2 \end{pmatrix}$$

THAT'S TRUE.

IF WE WANTED TO, WE COULD ALSO REWRITE IT LIKE THIS:

$$\begin{pmatrix} y_1 \\ y_2 \\ 1 \end{pmatrix} = \begin{pmatrix} x_1 + b_1 \\ x_2 + b_2 \\ 1 \end{pmatrix} = \begin{pmatrix} 1 & 0 & b_1 \\ 0 & 1 & b_2 \\ 0 & 0 & 1 \end{pmatrix}\begin{pmatrix} x_1 \\ x_2 \\ 1 \end{pmatrix}$$

SO APPLYING THE SET OF RULES $\begin{cases} \text{Translate all } x_1 \text{ values by } b_1 \\ \text{Translate all } x_2 \text{ values by } b_2 \end{cases}$

ONTO AN ARBITRARY IMAGE IS BASICALLY THE SAME THING AS PASSING THE IMAGE THROUGH A LINEAR TRANSFORMATION IN R^3 EQUAL TO THE FOLLOWING MATRIX:

$$\begin{pmatrix} 1 & 0 & b_1 \\ 0 & 1 & b_2 \\ 0 & 0 & 1 \end{pmatrix}$$

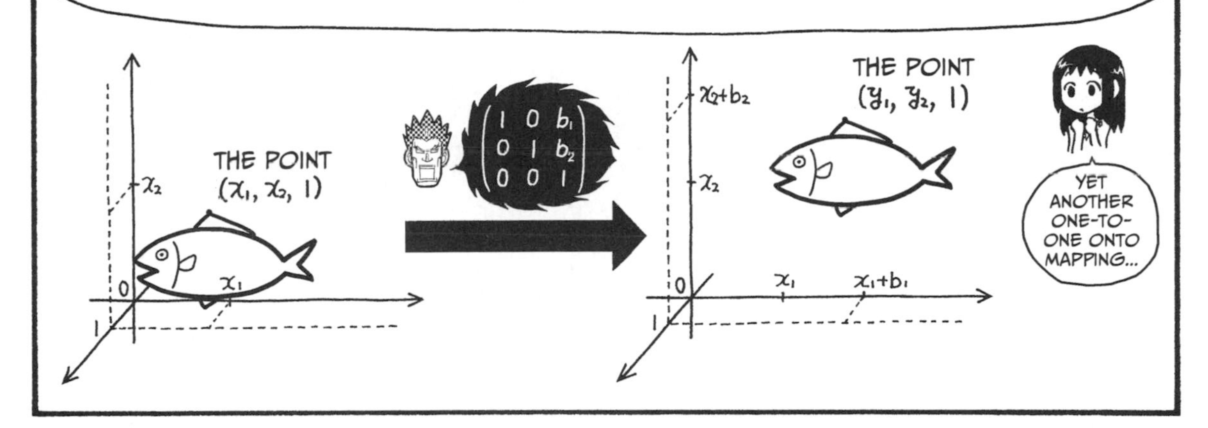

WE'D LIKE TO EXPRESS TRANSLATIONS IN THE SAME WAY AS ROTATIONS AND SCALE OPERATIONS, WITH

$$\begin{pmatrix} y_1 \\ y_2 \end{pmatrix} = \begin{pmatrix} a_{11} & a_{12} \\ a_{21} & a_{22} \end{pmatrix} \begin{pmatrix} x_1 \\ x_2 \end{pmatrix}$$

INSTEAD OF WITH

$$\begin{pmatrix} y_1 \\ y_2 \end{pmatrix} = \begin{pmatrix} a_{11} & a_{12} \\ a_{21} & a_{22} \end{pmatrix} \begin{pmatrix} x_1 \\ x_2 \end{pmatrix} + \begin{pmatrix} b_1 \\ b_2 \end{pmatrix}$$

THE FIRST FORMULA IS MORE PRACTICAL THAN THE SECOND, ESPECIALLY WHEN DEALING WITH COMPUTER GRAPHICS.

...EVEN ROTATIONS AND SCALING OPERATIONS.

NOT TOO DIFFERENT, I GUESS.

	CONVENTIONAL LINEAR TRANSFORMATIONS	LINEAR TRANSFORMATIONS USED BY COMPUTER GRAPHICS SYSTEMS
SCALING	$\begin{pmatrix} y_1 \\ y_2 \end{pmatrix} = \begin{pmatrix} \alpha & 0 \\ 0 & \beta \end{pmatrix} \begin{pmatrix} x_1 \\ x_2 \end{pmatrix}$	$\begin{pmatrix} y_1 \\ y_2 \\ 1 \end{pmatrix} = \begin{pmatrix} \alpha & 0 & 0 \\ 0 & \beta & 0 \\ 0 & 0 & 1 \end{pmatrix} \begin{pmatrix} x_1 \\ x_2 \\ 1 \end{pmatrix}$
ROTATION	$\begin{pmatrix} y_1 \\ y_2 \end{pmatrix} = \begin{pmatrix} \cos\theta & -\sin\theta \\ \sin\theta & \cos\theta \end{pmatrix} \begin{pmatrix} x_1 \\ x_2 \end{pmatrix}$	$\begin{pmatrix} y_1 \\ y_2 \\ 1 \end{pmatrix} = \begin{pmatrix} \cos\theta & -\sin\theta & 0 \\ \sin\theta & \cos\theta & 0 \\ 0 & 0 & 1 \end{pmatrix} \begin{pmatrix} x_1 \\ x_2 \\ 1 \end{pmatrix}$
TRANSLATION	$\begin{pmatrix} y_1 \\ y_2 \end{pmatrix} = \begin{pmatrix} 1 & 0 \\ 0 & 1 \end{pmatrix} \begin{pmatrix} x_1 \\ x_2 \end{pmatrix} + \begin{pmatrix} b_1 \\ b_2 \end{pmatrix}^{*}$	$\begin{pmatrix} y_1 \\ y_2 \\ 1 \end{pmatrix} = \begin{pmatrix} 1 & 0 & b_1 \\ 0 & 1 & b_2 \\ 0 & 0 & 1 \end{pmatrix} \begin{pmatrix} x_1 \\ x_2 \\ 1 \end{pmatrix}$

* NOTE: THIS ONE ISN'T ACTUALLY A LINEAR TRANSFORMATION. YOU CAN VERIFY THIS BY SETTING b_1 AND b_2 TO 1 AND CHECKING THAT BOTH LINEAR TRANSFORMATION CONDITIONS FAIL.

3-D PROJECTION
NEXT WE'LL VERY BRIEFLY TALK ABOUT A 3-D PROJECTION TECHNIQUE CALLED *PERSPECTIVE PROJECTION.*

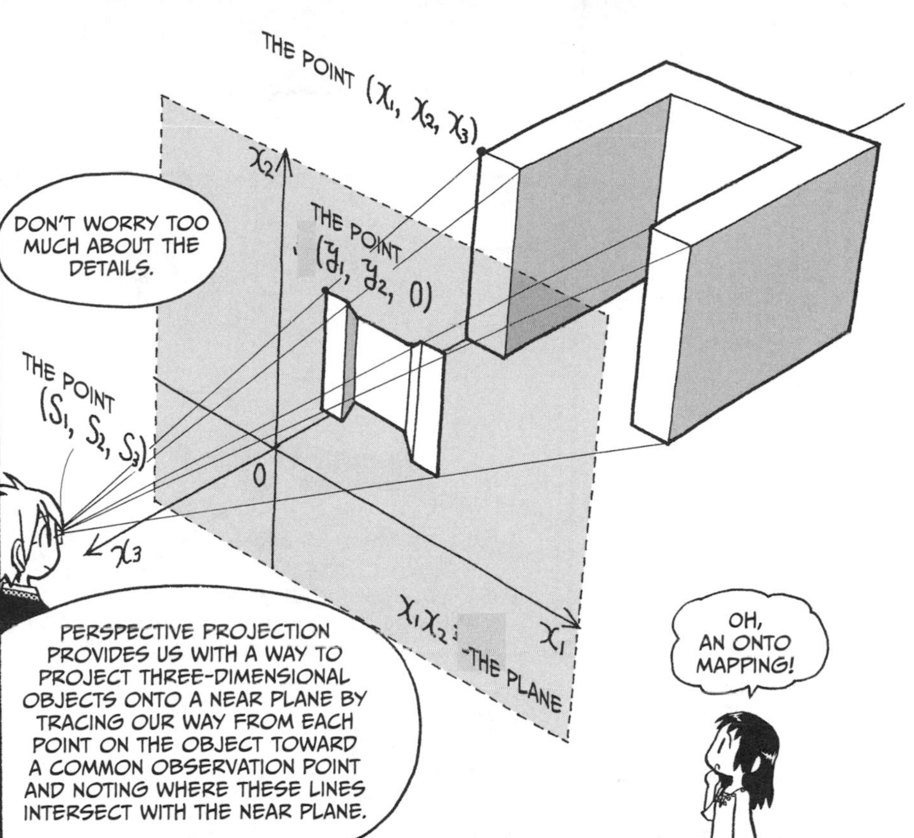
DON'T WORRY TOO MUCH ABOUT THE DETAILS.
THE POINT (x_1, x_2, x_3)
THE POINT $(y_1, y_2, 0)$
THE POINT (s_1, s_2, s_3)
x_2
0
x_3
x_1
x_1x_2-THE PLANE
PERSPECTIVE PROJECTION PROVIDES US WITH A WAY TO PROJECT THREE-DIMENSIONAL OBJECTS ONTO A NEAR PLANE BY TRACING OUR WAY FROM EACH POINT ON THE OBJECT TOWARD A COMMON OBSERVATION POINT AND NOTING WHERE THESE LINES INTERSECT WITH THE NEAR PLANE.
OH, AN ONTO MAPPING!

THE MATH IS A BIT MORE COMPLEX THAN WHAT WE'VE SEEN SO FAR.
SO I'LL CHEAT A LITTLE BIT AND SKIP RIGHT TO THE END!

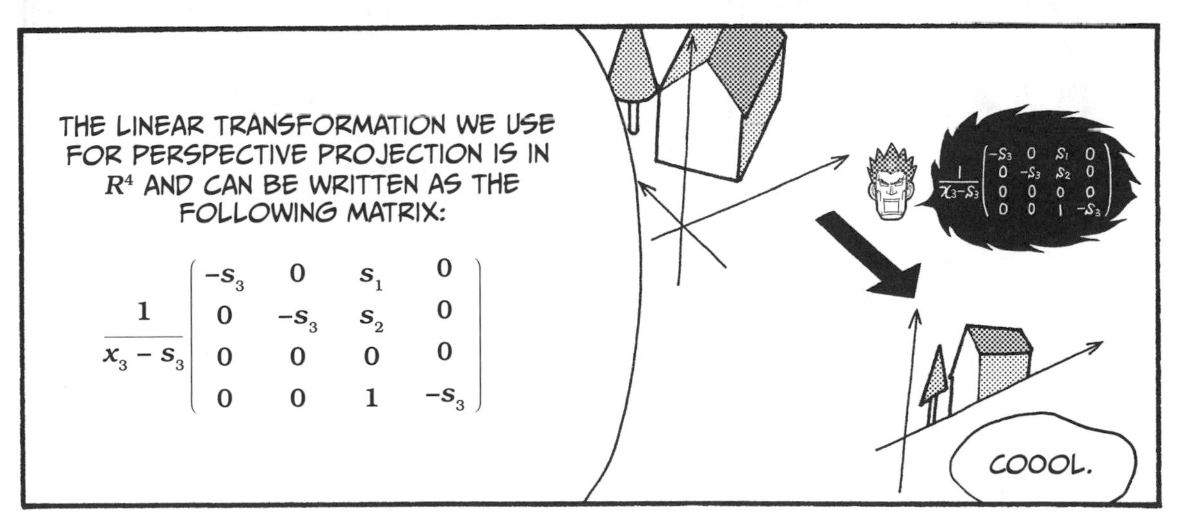
THE LINEAR TRANSFORMATION WE USE FOR PERSPECTIVE PROJECTION IS IN R^4 AND CAN BE WRITTEN AS THE FOLLOWING MATRIX:
$$\frac{1}{x_3 - s_3}\begin{pmatrix} -s_3 & 0 & s_1 & 0 \\ 0 & -s_3 & s_2 & 0 \\ 0 & 0 & 0 & 0 \\ 0 & 0 & 1 & -s_3 \end{pmatrix}$$
COOOL.

AND THAT'S WHAT TRANSFORMATIONS ARE ALL ABOUT!
SO MUCH TO LEARN...

YEAH...BUT THAT'S ENOUGH FOR TODAY, I THINK.
WE'LL BE TALKING ABOUT EIGENVECTORS AND EIGENVALUES IN OUR NEXT AND FINAL LESSON.

FINAL LESSON? SO SOON?
DON'T WORRY, WE'LL COVER ALL THE IMPORTANT TOPICS.

HEHE, WHY WOULD I WORRY? YOU'RE SUCH A GOOD TEACHER.
BEAM

YOU SHOULDN'T WORRY EITHER, YOU KNOW.
HM?
ABOUT THE MATCH.

OH, YOU HEARD?
YEAH, MY BROTHER TOLD ME.

ZZZZZIP
HEH. THANKS. I'M GOING TO THE GYM AFTER THIS, ACTUALLY. I HOPE I DON'T LOSE TOO BADLY...

DON'T SAY THAT!

YOU'VE GOT TO STAY POSITIVE!
TH- THANKS...
I KNOW YOU CAN DO IT.
I'LL DO MY BEST!

SOME PRELIMINARY TIPS

Before we dive into kernel, rank, and the other advanced topics we're going to cover in the remainder of this chapter, there's a little mathematical trick that you may find handy while working some of these problems out.

The equation

$$\begin{pmatrix} y_1 \\ y_2 \\ \vdots \\ y_m \end{pmatrix} = \begin{pmatrix} a_{11} & a_{12} & \cdots & a_{1n} \\ a_{21} & a_{22} & \cdots & a_{2n} \\ \vdots & \vdots & \ddots & \vdots \\ a_{m1} & a_{m2} & \cdots & a_{mn} \end{pmatrix} \begin{pmatrix} x_1 \\ x_2 \\ \vdots \\ x_n \end{pmatrix}$$

can be rewritten like this:

$$\begin{pmatrix} y_1 \\ y_2 \\ \vdots \\ y_m \end{pmatrix} = \begin{pmatrix} a_{11} & a_{12} & \cdots & a_{1n} \\ a_{21} & a_{22} & \cdots & a_{2n} \\ \vdots & \vdots & \ddots & \vdots \\ a_{m1} & a_{m2} & \cdots & a_{mn} \end{pmatrix} \begin{pmatrix} x_1 \\ x_2 \\ \vdots \\ x_n \end{pmatrix}$$

$$= \begin{pmatrix} a_{11} & a_{12} & \cdots & a_{1n} \\ a_{21} & a_{22} & \cdots & a_{2n} \\ \vdots & \vdots & \ddots & \vdots \\ a_{m1} & a_{m2} & \cdots & a_{mn} \end{pmatrix} \left[x_1 \begin{pmatrix} 1 \\ 0 \\ \vdots \\ 0 \end{pmatrix} + x_2 \begin{pmatrix} 0 \\ 1 \\ \vdots \\ 0 \end{pmatrix} + \ldots + x_n \begin{pmatrix} 0 \\ 0 \\ \vdots \\ 1 \end{pmatrix} \right]$$

$$= x_1 \begin{pmatrix} a_{11} \\ a_{21} \\ \vdots \\ a_{m1} \end{pmatrix} + x_2 \begin{pmatrix} a_{12} \\ a_{22} \\ \vdots \\ a_{m2} \end{pmatrix} + \ldots + x_n \begin{pmatrix} a_{1n} \\ a_{2n} \\ \vdots \\ a_{mn} \end{pmatrix}$$

As you can see, the product of the matrix M and the vector $\boldsymbol{x}$ can be viewed as a linear combination of the columns of M with the entries of $\boldsymbol{x}$ as the weights.

Also note that the function f referred to throughout this chapter is the linear transformation from R^n to R^m corresponding to the following $m \times n$ matrix:

$$\begin{pmatrix} a_{11} & a_{12} & \cdots & a_{1n} \\ a_{21} & a_{22} & \cdots & a_{2n} \\ \vdots & \vdots & \ddots & \vdots \\ a_{m1} & a_{m2} & \cdots & a_{mn} \end{pmatrix}$$

KERNEL, IMAGE, AND THE DIMENSION THEOREM FOR LINEAR TRANSFORMATIONS

The set of vectors whose images are the zero vector, that is

$$\left\{ \begin{pmatrix} x_1 \\ x_2 \\ \vdots \\ x_n \end{pmatrix} \middle| \begin{pmatrix} 0 \\ 0 \\ \vdots \\ 0 \end{pmatrix} = \begin{pmatrix} a_{11} & a_{12} & \cdots & a_{1n} \\ a_{21} & a_{22} & \cdots & a_{2n} \\ \vdots & \vdots & \ddots & \vdots \\ a_{m1} & a_{m2} & \cdots & a_{mn} \end{pmatrix} \begin{pmatrix} x_1 \\ x_2 \\ \vdots \\ x_n \end{pmatrix} \right\}$$

is called the *kernel* of the linear transformation f and is written Ker f.

The *image* of f (written Im f) is also important in this context. The image of f is equal to the set of vectors that is made up of all of the possible output values of f, as you can see in the following relation:

$$\left\{ \begin{pmatrix} y_1 \\ y_2 \\ \vdots \\ y_m \end{pmatrix} \middle| \begin{pmatrix} y_1 \\ y_2 \\ \vdots \\ y_m \end{pmatrix} = \begin{pmatrix} a_{11} & a_{12} & \cdots & a_{1n} \\ a_{21} & a_{22} & \cdots & a_{2n} \\ \vdots & \vdots & \ddots & \vdots \\ a_{m1} & a_{m2} & \cdots & a_{mn} \end{pmatrix} \begin{pmatrix} x_1 \\ x_2 \\ \vdots \\ x_n \end{pmatrix} \right\}$$

(This is a more formal definition of image than what we saw in Chapter 2, but the concept is the same.)

An important observation is that Ker f is a subspace of R^n and Im f is a subspace of R^m. The *dimension theorem for linear transformations* further explores this observation by defining a relationship between the two:

$$\dim \operatorname{Ker} f + \dim \operatorname{Im} f = n$$

Note that the n above is equal to the first vector space's dimension (dim R^n).*

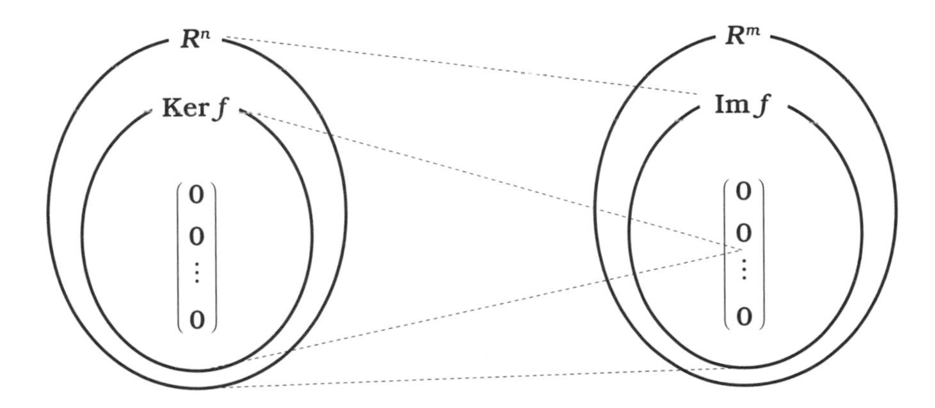

* If you need a refresher on the concept of dimension, see "Basis and Dimension" on page 156.

EXAMPLE 1

Suppose that f is a linear transformation from R^2 to R^2 equal to the matrix $\begin{pmatrix} 3 & 1 \\ 1 & 2 \end{pmatrix}$. Then:

$$\begin{cases} \operatorname{Ker} f = \left\{ \begin{pmatrix} x_1 \\ x_2 \end{pmatrix} \middle| \begin{pmatrix} 0 \\ 0 \end{pmatrix} = \begin{pmatrix} 3 & 1 \\ 1 & 2 \end{pmatrix} \begin{pmatrix} x_1 \\ x_2 \end{pmatrix} \right\} = \left\{ \begin{pmatrix} x_1 \\ x_2 \end{pmatrix} \middle| \begin{pmatrix} 0 \\ 0 \end{pmatrix} = x_1 \begin{pmatrix} 3 \\ 1 \end{pmatrix} + x_2 \begin{pmatrix} 1 \\ 2 \end{pmatrix} \right\} = \left\{ \begin{pmatrix} 0 \\ 0 \end{pmatrix} \right\} \\ \operatorname{Im} f = \left\{ \begin{pmatrix} y_1 \\ y_2 \end{pmatrix} \middle| \begin{pmatrix} y_1 \\ y_2 \end{pmatrix} = \begin{pmatrix} 3 & 1 \\ 1 & 2 \end{pmatrix} \begin{pmatrix} x_1 \\ x_2 \end{pmatrix} \right\} = \left\{ \begin{pmatrix} y_1 \\ y_2 \end{pmatrix} \middle| \begin{pmatrix} y_1 \\ y_2 \end{pmatrix} = x_1 \begin{pmatrix} 3 \\ 1 \end{pmatrix} + x_2 \begin{pmatrix} 1 \\ 2 \end{pmatrix} \right\} = R^2 \end{cases}$$

And: $\begin{cases} n & = 2 \\ \dim \operatorname{Ker} f & = 0 \\ \dim \operatorname{Im} f & = 2 \end{cases}$

EXAMPLE 2

Suppose that f is a linear transformation from R^2 to R^2 equal to the matrix $\begin{pmatrix} 3 & 6 \\ 1 & 2 \end{pmatrix}$. Then:

$$\begin{cases} \operatorname{Ker} f = \left\{ \begin{pmatrix} x_1 \\ x_2 \end{pmatrix} \middle| \begin{pmatrix} 0 \\ 0 \end{pmatrix} = \begin{pmatrix} 3 & 6 \\ 1 & 2 \end{pmatrix} \begin{pmatrix} x_1 \\ x_2 \end{pmatrix} \right\} = \left\{ \begin{pmatrix} x_1 \\ x_2 \end{pmatrix} \middle| \begin{pmatrix} 0 \\ 0 \end{pmatrix} = [x_1 + 2x_2] \begin{pmatrix} 3 \\ 1 \end{pmatrix} \right\} \\ \qquad = \left\{ c \begin{pmatrix} -2 \\ 1 \end{pmatrix} \middle| \text{c is an arbitrary number} \right\} \\ \operatorname{Im} f = \left\{ \begin{pmatrix} y_1 \\ y_2 \end{pmatrix} \middle| \begin{pmatrix} y_1 \\ y_2 \end{pmatrix} = \begin{pmatrix} 3 & 6 \\ 1 & 2 \end{pmatrix} \begin{pmatrix} x_1 \\ x_2 \end{pmatrix} \right\} = \left\{ \begin{pmatrix} y_1 \\ y_2 \end{pmatrix} \middle| \begin{pmatrix} y_1 \\ y_2 \end{pmatrix} = [x_1 + 2x_2] \begin{pmatrix} 3 \\ 1 \end{pmatrix} \right\} \\ \qquad = \left\{ c \begin{pmatrix} 3 \\ 1 \end{pmatrix} \middle| \text{c is an arbitrary number} \right\} \end{cases}$$

And: $\begin{cases} n & = 2 \\ \dim \operatorname{Ker} f & = 1 \\ \dim \operatorname{Im} f & = 1 \end{cases}$

EXAMPLE 3

Suppose f is a linear transformation from R^2 to R^3 equal to the 3×2 matrix $\begin{pmatrix} 1 & 0 \\ 0 & 1 \\ 0 & 0 \end{pmatrix}$. Then:

$$\begin{cases} \operatorname{Ker} f = \left\{ \begin{pmatrix} x_1 \\ x_2 \end{pmatrix} \middle| \begin{pmatrix} 0 \\ 0 \\ 0 \end{pmatrix} = \begin{pmatrix} 1 & 0 \\ 0 & 1 \\ 0 & 0 \end{pmatrix} \begin{pmatrix} x_1 \\ x_2 \end{pmatrix} \right\} = \left\{ \begin{pmatrix} x_1 \\ x_2 \end{pmatrix} \middle| \begin{pmatrix} 0 \\ 0 \\ 0 \end{pmatrix} = x_1 \begin{pmatrix} 1 \\ 0 \\ 0 \end{pmatrix} + x_2 \begin{pmatrix} 0 \\ 1 \\ 0 \end{pmatrix} \right\} = \left\{ \begin{pmatrix} 0 \\ 0 \end{pmatrix} \right\} \\ \operatorname{Im} f = \left\{ \begin{pmatrix} y_1 \\ y_2 \\ y_3 \end{pmatrix} \middle| \begin{pmatrix} y_1 \\ y_2 \\ y_3 \end{pmatrix} = \begin{pmatrix} 1 & 0 \\ 0 & 1 \\ 0 & 0 \end{pmatrix} \begin{pmatrix} x_1 \\ x_2 \end{pmatrix} \right\} = \left\{ \begin{pmatrix} y_1 \\ y_2 \\ y_3 \end{pmatrix} \middle| \begin{pmatrix} y_1 \\ y_2 \\ y_3 \end{pmatrix} = x_1 \begin{pmatrix} 1 \\ 0 \\ 0 \end{pmatrix} + x_2 \begin{pmatrix} 0 \\ 1 \\ 0 \end{pmatrix} \right\} \\ \qquad = \left\{ c_1 \begin{pmatrix} 1 \\ 0 \\ 0 \end{pmatrix} + c_2 \begin{pmatrix} 0 \\ 1 \\ 0 \end{pmatrix} \middle| \text{c_1 and c_2 are arbitrary numbers} \right\} \end{cases}$$

And: $\begin{cases} n & = 2 \\ \dim \operatorname{Ker} f & = 0 \\ \dim \operatorname{Im} f & = 2 \end{cases}$

EXAMPLE 4

Suppose that f is a linear transformation from R^4 to R^2 equal to the 2×4 matrix $\begin{pmatrix} 1 & 0 & 3 & 1 \\ 0 & 1 & 1 & 2 \end{pmatrix}$. Then:

$$\begin{cases} \operatorname{Ker} f = \left\{ \begin{pmatrix} x_1 \\ x_2 \\ x_3 \\ x_4 \end{pmatrix} \middle| \begin{pmatrix} 0 \\ 0 \end{pmatrix} = \begin{pmatrix} 1 & 0 & 3 & 1 \\ 0 & 1 & 1 & 2 \end{pmatrix} \begin{pmatrix} x_1 \\ x_2 \\ x_3 \\ x_4 \end{pmatrix} \right\} \\ \quad = \left\{ \begin{pmatrix} x_1 \\ x_2 \\ x_3 \\ x_4 \end{pmatrix} \middle| \begin{pmatrix} 0 \\ 0 \end{pmatrix} = x_1 \begin{pmatrix} 1 \\ 0 \end{pmatrix} + x_2 \begin{pmatrix} 0 \\ 1 \end{pmatrix} + x_3 \begin{pmatrix} 3 \\ 1 \end{pmatrix} + x_4 \begin{pmatrix} 1 \\ 2 \end{pmatrix} \right\} \\ \quad = \left\{ \begin{pmatrix} x_1 \\ x_2 \\ x_3 \\ x_4 \end{pmatrix} \middle| x_1 + 3x_3 + x_4 = 0,\ x_2 + x_3 + 2x_4 = 0 \right\} \\ \quad = \left\{ c_1 \begin{pmatrix} -3 \\ -1 \\ 1 \\ 0 \end{pmatrix} + c_2 \begin{pmatrix} -1 \\ -2 \\ 0 \\ 1 \end{pmatrix} \middle| \text{c_1 and c_2 are arbitrary numbers} \right\} \\ \operatorname{Im} f = \left\{ \begin{pmatrix} y_1 \\ y_2 \end{pmatrix} \middle| \begin{pmatrix} y_1 \\ y_2 \end{pmatrix} = \begin{pmatrix} 1 & 0 & 3 & 1 \\ 0 & 1 & 1 & 2 \end{pmatrix} \begin{pmatrix} x_1 \\ x_2 \\ x_3 \\ x_4 \end{pmatrix} \right\} \\ \quad = \left\{ \begin{pmatrix} y_1 \\ y_2 \end{pmatrix} \middle| \begin{pmatrix} y_1 \\ y_2 \end{pmatrix} = x_1 \begin{pmatrix} 1 \\ 0 \end{pmatrix} + x_2 \begin{pmatrix} 0 \\ 1 \end{pmatrix} + x_3 \begin{pmatrix} 3 \\ 1 \end{pmatrix} + x_4 \begin{pmatrix} 1 \\ 2 \end{pmatrix} \right\} = R^2 \end{cases}$$

And: $\begin{cases} n & = 4 \\ \dim \operatorname{Ker} f & = 2 \\ \dim \operatorname{Im} f & = 2 \end{cases}$

RANK

The number of linearly independent vectors among the columns of the matrix M (which is also the dimension of the R^m subspace $\mathrm{Im}\, f$) is called the *rank* of M, and it is written like this: rank M.

EXAMPLE 1

The linear system of equations $\begin{cases} 3x_1 + 1x_2 = y_1 \\ 1x_1 + 2x_2 = y_2 \end{cases}$, that is $\begin{pmatrix} y_1 \\ y_2 \end{pmatrix} = \begin{pmatrix} 3x_1 + 1x_2 \\ 1x_1 + 2x_2 \end{pmatrix}$,

can be rewritten as follows: $\begin{pmatrix} y_1 \\ y_2 \end{pmatrix} = \begin{pmatrix} 3x_1 + 1x_2 \\ 1x_1 + 2x_2 \end{pmatrix} = \begin{pmatrix} 3 & 1 \\ 1 & 2 \end{pmatrix}\begin{pmatrix} x_1 \\ x_2 \end{pmatrix} = x_1\begin{pmatrix} 3 \\ 1 \end{pmatrix} + x_2\begin{pmatrix} 1 \\ 2 \end{pmatrix}$

The two vectors $\begin{pmatrix} 3 \\ 1 \end{pmatrix}$ and $\begin{pmatrix} 1 \\ 2 \end{pmatrix}$ are linearly independent, as can be seen on pages 133 and 135, so the rank of $\begin{pmatrix} 3 & 1 \\ 1 & 2 \end{pmatrix}$ is 2.

Also note that $\det\begin{pmatrix} 3 & 1 \\ 1 & 2 \end{pmatrix} = 3 \cdot 2 - 1 \cdot 1 = 5 \neq 0$.

EXAMPLE 2

The linear system of equations $\begin{cases} 3x_1 + 6x_2 = y_1 \\ 1x_1 + 2x_2 = y_2 \end{cases}$, that is $\begin{pmatrix} y_1 \\ y_2 \end{pmatrix} = \begin{pmatrix} 3x_1 + 6x_2 \\ 1x_1 + 2x_2 \end{pmatrix}$,

can be rewritten as follows:

$$\begin{pmatrix} y_1 \\ y_2 \end{pmatrix} = \begin{pmatrix} 3x_1 + 6x_2 \\ 1x_1 + 2x_2 \end{pmatrix} = \begin{pmatrix} 3 & 6 \\ 1 & 2 \end{pmatrix}\begin{pmatrix} x_1 \\ x_2 \end{pmatrix} = x_1\begin{pmatrix} 3 \\ 1 \end{pmatrix} + x_2\begin{pmatrix} 6 \\ 2 \end{pmatrix}$$

$$= x_1\begin{pmatrix} 3 \\ 1 \end{pmatrix} + 2x_2\begin{pmatrix} 3 \\ 1 \end{pmatrix}$$

$$= [x_1 + 2x_2]\begin{pmatrix} 3 \\ 1 \end{pmatrix}$$

So the rank of $\begin{pmatrix} 3 & 6 \\ 1 & 2 \end{pmatrix}$ is 1.

Also note that $\det\begin{pmatrix} 3 & 6 \\ 1 & 2 \end{pmatrix} = 3 \cdot 2 - 6 \cdot 1 = 0$.

EXAMPLE 3

The linear system of equations $\begin{cases} 1x_1 + 0x_2 = y_1 \\ 0x_1 + 1x_2 = y_2 \\ 0x_1 + 0x_2 = y_3 \end{cases}$, that is $\begin{pmatrix} y_1 \\ y_2 \\ y_3 \end{pmatrix} = \begin{pmatrix} 1x_1 + 0x_2 \\ 0x_1 + 1x_2 \\ 0x_1 + 0x_2 \end{pmatrix}$,

can be rewritten as: $\begin{pmatrix} y_1 \\ y_2 \\ y_3 \end{pmatrix} = \begin{pmatrix} 1x_1 + 0x_2 \\ 0x_1 + 1x_2 \\ 0x_1 + 0x_2 \end{pmatrix} = \begin{pmatrix} 1 & 0 \\ 0 & 1 \\ 0 & 0 \end{pmatrix} \begin{pmatrix} x_1 \\ x_2 \end{pmatrix} = x_1 \begin{pmatrix} 1 \\ 0 \\ 0 \end{pmatrix} + x_2 \begin{pmatrix} 0 \\ 1 \\ 0 \end{pmatrix}$

The two vectors $\begin{pmatrix} 1 \\ 0 \\ 0 \end{pmatrix}$ and $\begin{pmatrix} 0 \\ 1 \\ 0 \end{pmatrix}$ are linearly independent, as we discovered on page 137, so the rank of $\begin{pmatrix} 1 & 0 \\ 0 & 1 \\ 0 & 0 \end{pmatrix}$ is 2.

The system could also be rewritten like this:

$$\begin{pmatrix} y_1 \\ y_2 \\ y_3 \end{pmatrix} = \begin{pmatrix} 1x_1 + 0x_2 \\ 0x_1 + 1x_2 \\ 0x_1 + 0x_2 \end{pmatrix} = \begin{pmatrix} 1 & 0 & 0 \\ 0 & 1 & 0 \\ 0 & 0 & 0 \end{pmatrix} \begin{pmatrix} x_1 \\ x_2 \\ x_3 \end{pmatrix}$$

Note that $\det \begin{pmatrix} 1 & 0 & 0 \\ 0 & 1 & 0 \\ 0 & 0 & 0 \end{pmatrix} = 0$.

EXAMPLE 4

The linear system of equations $\begin{cases} 1x_1 + 0x_2 + 3x_3 + 1x_4 = y_1 \\ 0x_1 + 1x_2 + 1x_3 + 2x_4 = y_2 \end{cases}$, that is

$\begin{pmatrix} y_1 \\ y_2 \end{pmatrix} = \begin{pmatrix} 1x_1 + 0x_2 + 3x_3 + 1x_4 \\ 0x_1 + 1x_2 + 1x_3 + 2x_4 \end{pmatrix}$, can be rewritten as follows:

$$\begin{pmatrix} y_1 \\ y_2 \end{pmatrix} = \begin{pmatrix} 1x_1 + 0x_2 + 3x_3 + 1x_4 \\ 0x_1 + 1x_2 + 1x_3 + 2x_4 \end{pmatrix} = \begin{pmatrix} 1 & 0 & 3 & 1 \\ 0 & 1 & 1 & 2 \end{pmatrix} \begin{pmatrix} x_1 \\ x_2 \\ x_3 \\ x_4 \end{pmatrix}$$

$$= x_1 \begin{pmatrix} 1 \\ 0 \end{pmatrix} + x_2 \begin{pmatrix} 0 \\ 1 \end{pmatrix} + x_3 \begin{pmatrix} 3 \\ 1 \end{pmatrix} + x_4 \begin{pmatrix} 1 \\ 2 \end{pmatrix}$$

The rank of $\begin{pmatrix} 1 & 0 & 3 & 1 \\ 0 & 1 & 1 & 2 \end{pmatrix}$ is equal to 2, as we'll see on page 203.

The system could also be rewritten like this:

$$\begin{pmatrix} y_1 \\ y_2 \\ y_3 \\ y_4 \end{pmatrix} = \begin{pmatrix} 1x_1 + 0x_2 + 3x_3 + 1x_4 \\ 0x_1 + 1x_2 + 1x_3 + 2x_4 \\ 0 \\ 0 \end{pmatrix} = \begin{pmatrix} 1 & 0 & 3 & 1 \\ 0 & 1 & 1 & 2 \\ 0 & 0 & 0 & 0 \\ 0 & 0 & 0 & 0 \end{pmatrix} \begin{pmatrix} x_1 \\ x_2 \\ x_3 \\ x_4 \end{pmatrix}$$

Note that $\det \begin{pmatrix} 1 & 0 & 3 & 1 \\ 0 & 1 & 1 & 2 \\ 0 & 0 & 0 & 0 \\ 0 & 0 & 0 & 0 \end{pmatrix} = 0$.

The four examples seem to point to the fact that

$$\det \begin{pmatrix} a_{11} & a_{12} & \cdots & a_{1n} \\ a_{21} & a_{22} & \cdots & a_{2n} \\ \vdots & \vdots & \ddots & \vdots \\ a_{n1} & a_{n2} & \cdots & a_{nn} \end{pmatrix} = 0 \text{ is the same as rank } \begin{pmatrix} a_{11} & a_{12} & \cdots & a_{1n} \\ a_{21} & a_{22} & \cdots & a_{2n} \\ \vdots & \vdots & \ddots & \vdots \\ a_{n1} & a_{n2} & \cdots & a_{nn} \end{pmatrix} \neq n.$$

This is indeed so, but no formal proof will be given in this book.

CALCULATING THE RANK OF A MATRIX

So far, we've only dealt with matrices where the rank was immediately apparent or where we had previously figured out how many linearly independent vectors made up the columns of that matrix. Though this might seem like "cheating" at first, these techniques can actually be very useful for calculating ranks in practice.

For example, take a look at the following matrix:

$$\begin{pmatrix} 1 & 4 & 4 \\ 2 & 5 & 8 \\ 3 & 6 & 12 \end{pmatrix}$$

It's immediately clear that the third column of this matrix is equal to the first column times 4. This leaves two linearly independent vectors (the first two columns), which means this matrix has a rank of 2.

Now look at this matrix:

$$\begin{pmatrix} 1 & 0 \\ 0 & 3 \\ 0 & 5 \end{pmatrix}$$

It should be obvious right from the start that these vectors form a linearly independent set, so we know that the rank of this matrix is also 2.

Of course there are times when this method will fail you and you won't be able to tell the rank of a matrix just by eyeballing it. In those cases, you'll have to buckle down and actually calculate the rank. But don't worry, it's not too hard!

First we'll explain the PROBLEM, then we'll establish a good WAY OF THINKING, and then finally we'll tackle the SOLUTION.

PROBLEM

Calculate the rank of the following 2×4 matrix:

$$\begin{pmatrix} 1 & 0 & 3 & 1 \\ 0 & 1 & 1 & 2 \end{pmatrix}$$

WAY OF THINKING

Before we can solve this problem, we need to learn a little bit about elementary matrices. An *elementary matrix* is created by starting with an identity matrix and performing exactly one of the elementary row operations used for Gaussian elimination (see Chapter 4). The resulting matrices can then be multiplied with any arbitrary matrix in such a way that the number of linearly independent columns becomes obvious.

With this information under our belts, we can state the following four useful facts about an arbitrary matrix A:

$$\begin{pmatrix} a_{11} & a_{12} & \cdots & a_{1n} \\ a_{21} & a_{22} & \cdots & a_{2n} \\ \vdots & \vdots & \ddots & \vdots \\ a_{m1} & a_{m2} & \cdots & a_{mn} \end{pmatrix}$$

FACT 1

Multiplying the elementary matrix

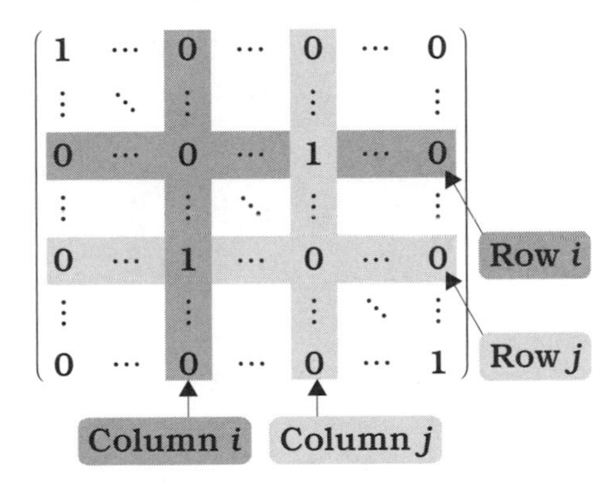

to the left of an arbitrary matrix A will switch rows i and j in A.

If we multiply the matrix to the right of A, then the columns will switch places in A instead.

- Example 1 (Rows 1 and 4 are switched.)

$$\begin{pmatrix} 0 & 0 & 0 & 1 \\ 0 & 1 & 0 & 0 \\ 0 & 0 & 1 & 0 \\ 1 & 0 & 0 & 0 \end{pmatrix} \begin{pmatrix} a_{11} & a_{12} & a_{13} \\ a_{21} & a_{22} & a_{23} \\ a_{31} & a_{32} & a_{33} \\ a_{41} & a_{42} & a_{43} \end{pmatrix}$$

$$= \begin{pmatrix} 0 \cdot a_{11} + 0 \cdot a_{21} + 0 \cdot a_{31} + 1 \cdot a_{41} & 0 \cdot a_{12} + 0 \cdot a_{22} + 0 \cdot a_{32} + 1 \cdot a_{42} & 0 \cdot a_{13} + 0 \cdot a_{23} + 0 \cdot a_{33} + 1 \cdot a_{43} \\ 0 \cdot a_{11} + 1 \cdot a_{21} + 0 \cdot a_{31} + 0 \cdot a_{41} & 0 \cdot a_{12} + 1 \cdot a_{22} + 0 \cdot a_{32} + 0 \cdot a_{42} & 0 \cdot a_{13} + 1 \cdot a_{23} + 0 \cdot a_{33} + 0 \cdot a_{43} \\ 0 \cdot a_{11} + 0 \cdot a_{21} + 1 \cdot a_{31} + 0 \cdot a_{41} & 0 \cdot a_{12} + 0 \cdot a_{22} + 1 \cdot a_{32} + 0 \cdot a_{42} & 0 \cdot a_{13} + 0 \cdot a_{23} + 1 \cdot a_{33} + 0 \cdot a_{43} \\ 1 \cdot a_{11} + 0 \cdot a_{21} + 0 \cdot a_{31} + 0 \cdot a_{41} & 1 \cdot a_{12} + 0 \cdot a_{22} + 0 \cdot a_{32} + 0 \cdot a_{42} & 1 \cdot a_{13} + 0 \cdot a_{23} + 0 \cdot a_{33} + 0 \cdot a_{43} \end{pmatrix}$$

$$= \begin{pmatrix} a_{41} & a_{42} & a_{43} \\ a_{21} & a_{22} & a_{23} \\ a_{31} & a_{32} & a_{33} \\ a_{11} & a_{12} & a_{13} \end{pmatrix}$$

- Example 2 (Columns 1 and 3 are switched.)

$$\begin{pmatrix} a_{11} & a_{12} & a_{13} \\ a_{21} & a_{22} & a_{23} \\ a_{31} & a_{32} & a_{33} \\ a_{41} & a_{42} & a_{43} \end{pmatrix} \begin{pmatrix} 0 & 0 & 1 \\ 0 & 1 & 0 \\ 1 & 0 & 0 \end{pmatrix}$$

$$= \begin{pmatrix} a_{11}\cdot 0 + a_{12}\cdot 0 + a_{13}\cdot 1 & a_{11}\cdot 0 + a_{12}\cdot 1 + a_{13}\cdot 0 & a_{11}\cdot 1 + a_{12}\cdot 0 + a_{13}\cdot 0 \\ a_{21}\cdot 0 + a_{22}\cdot 0 + a_{23}\cdot 1 & a_{21}\cdot 0 + a_{22}\cdot 1 + a_{23}\cdot 0 & a_{21}\cdot 1 + a_{22}\cdot 0 + a_{23}\cdot 0 \\ a_{31}\cdot 0 + a_{32}\cdot 0 + a_{33}\cdot 1 & a_{31}\cdot 0 + a_{32}\cdot 1 + a_{33}\cdot 0 & a_{31}\cdot 1 + a_{32}\cdot 0 + a_{33}\cdot 0 \\ a_{41}\cdot 0 + a_{42}\cdot 0 + a_{43}\cdot 1 & a_{41}\cdot 0 + a_{42}\cdot 1 + a_{43}\cdot 0 & a_{41}\cdot 1 + a_{42}\cdot 0 + a_{43}\cdot 0 \end{pmatrix}$$

$$= \begin{pmatrix} a_{13} & a_{12} & a_{11} \\ a_{23} & a_{22} & a_{21} \\ a_{33} & a_{32} & a_{31} \\ a_{43} & a_{42} & a_{41} \end{pmatrix}$$

FACT 2

Multiplying the elementary matrix

$$\begin{pmatrix} 1 & \cdots & 0 & \cdots & 0 \\ \vdots & \ddots & \vdots & & \vdots \\ 0 & \cdots & k & \cdots & 0 \\ \vdots & & \vdots & \ddots & \vdots \\ 0 & \cdots & 0 & \cdots & 1 \end{pmatrix}$$

Row i

Column i

to the left of an arbitrary matrix A will multiply the ith row in A by k.

Multiplying the matrix to the right side of A will multiply the ith column in A by k instead.

- Example 1 (Row 3 is multiplied by k.)

$$\begin{pmatrix} 1 & 0 & 0 & 0 \\ 0 & 1 & 0 & 0 \\ 0 & 0 & k & 0 \\ 0 & 0 & 0 & 1 \end{pmatrix} \begin{pmatrix} a_{11} & a_{12} & a_{13} \\ a_{21} & a_{22} & a_{23} \\ a_{31} & a_{32} & a_{33} \\ a_{41} & a_{42} & a_{43} \end{pmatrix}$$

$$= \begin{pmatrix} 1 \cdot a_{11} + 0 \cdot a_{21} + 0 \cdot a_{31} + 0 \cdot a_{41} & 1 \cdot a_{12} + 0 \cdot a_{22} + 0 \cdot a_{32} + 0 \cdot a_{42} & 1 \cdot a_{13} + 0 \cdot a_{23} + 0 \cdot a_{33} + 0 \cdot a_{43} \\ 0 \cdot a_{11} + 1 \cdot a_{21} + 0 \cdot a_{31} + 0 \cdot a_{41} & 0 \cdot a_{12} + 1 \cdot a_{22} + 0 \cdot a_{32} + 0 \cdot a_{42} & 0 \cdot a_{13} + 1 \cdot a_{23} + 0 \cdot a_{33} + 0 \cdot a_{43} \\ 0 \cdot a_{11} + 0 \cdot a_{21} + k \cdot a_{31} + 0 \cdot a_{41} & 0 \cdot a_{12} + 0 \cdot a_{22} + k \cdot a_{32} + 0 \cdot a_{42} & 0 \cdot a_{13} + 0 \cdot a_{23} + k \cdot a_{33} + 0 \cdot a_{43} \\ 0 \cdot a_{11} + 0 \cdot a_{21} + 0 \cdot a_{31} + 1 \cdot a_{41} & 0 \cdot a_{12} + 0 \cdot a_{22} + 0 \cdot a_{32} + 1 \cdot a_{42} & 0 \cdot a_{13} + 0 \cdot a_{23} + 0 \cdot a_{33} + 1 \cdot a_{43} \end{pmatrix}$$

$$= \begin{pmatrix} a_{11} & a_{12} & a_{13} \\ a_{21} & a_{22} & a_{23} \\ ka_{31} & ka_{32} & ka_{33} \\ a_{41} & a_{42} & a_{43} \end{pmatrix}$$

- Example 2 (Column 2 is multiplied by k.)

$$\begin{pmatrix} a_{11} & a_{12} & a_{13} \\ a_{21} & a_{22} & a_{23} \\ a_{31} & a_{32} & a_{33} \\ a_{41} & a_{42} & a_{43} \end{pmatrix} \begin{pmatrix} 1 & 0 & 0 \\ 0 & k & 0 \\ 0 & 0 & 1 \end{pmatrix}$$

$$= \begin{pmatrix} a_{11} \cdot 1 + a_{12} \cdot 0 + a_{13} \cdot 0 & a_{11} \cdot 0 + a_{12} \cdot k + a_{13} \cdot 0 & a_{11} \cdot 0 + a_{12} \cdot 0 + a_{13} \cdot 1 \\ a_{21} \cdot 1 + a_{22} \cdot 0 + a_{23} \cdot 0 & a_{21} \cdot 0 + a_{22} \cdot k + a_{23} \cdot 0 & a_{21} \cdot 0 + a_{22} \cdot 0 + a_{23} \cdot 1 \\ a_{31} \cdot 1 + a_{32} \cdot 0 + a_{33} \cdot 0 & a_{31} \cdot 0 + a_{32} \cdot k + a_{33} \cdot 0 & a_{31} \cdot 0 + a_{32} \cdot 0 + a_{33} \cdot 1 \\ a_{41} \cdot 1 + a_{42} \cdot 0 + a_{43} \cdot 0 & a_{41} \cdot 0 + a_{42} \cdot k + a_{43} \cdot 0 & a_{41} \cdot 0 + a_{42} \cdot 0 + a_{43} \cdot 1 \end{pmatrix}$$

$$= \begin{pmatrix} a_{11} & ka_{12} & a_{13} \\ a_{21} & ka_{22} & a_{23} \\ a_{31} & ka_{32} & a_{33} \\ a_{41} & ka_{42} & a_{43} \end{pmatrix}$$

FACT 3

Multiplying the elementary matrix

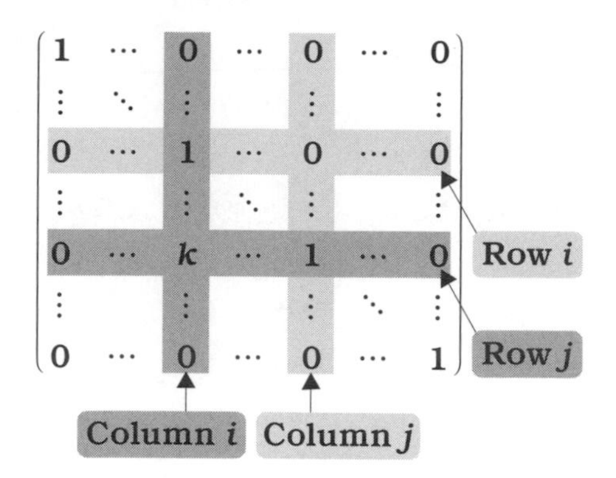

to the left of an arbitrary matrix A will add k times row i to row j in A.

Multiplying the matrix to the right side of A will add k times column j to column i instead.

- Example 1 (k times row 2 is added to row 4.)

$$\begin{pmatrix} 1 & 0 & 0 & 0 \\ 0 & 1 & 0 & 0 \\ 0 & 0 & 1 & 0 \\ 0 & k & 0 & 1 \end{pmatrix}\begin{pmatrix} a_{11} & a_{12} & a_{13} \\ a_{21} & a_{22} & a_{23} \\ a_{31} & a_{32} & a_{33} \\ a_{41} & a_{42} & a_{43} \end{pmatrix}$$

$$= \begin{pmatrix} 1 \cdot a_{11} + 0 \cdot a_{21} + 0 \cdot a_{31} + 0 \cdot a_{41} & 1 \cdot a_{12} + 0 \cdot a_{22} + 0 \cdot a_{32} + 0 \cdot a_{42} & 1 \cdot a_{13} + 0 \cdot a_{23} + 0 \cdot a_{33} + 0 \cdot a_{43} \\ 0 \cdot a_{11} + 1 \cdot a_{21} + 0 \cdot a_{31} + 0 \cdot a_{41} & 0 \cdot a_{12} + 1 \cdot a_{22} + 0 \cdot a_{32} + 0 \cdot a_{42} & 0 \cdot a_{13} + 1 \cdot a_{23} + 0 \cdot a_{33} + 0 \cdot a_{43} \\ 0 \cdot a_{11} + 0 \cdot a_{21} + 1 \cdot a_{31} + 0 \cdot a_{41} & 0 \cdot a_{12} + 0 \cdot a_{22} + 1 \cdot a_{32} + 0 \cdot a_{42} & 0 \cdot a_{13} + 0 \cdot a_{23} + 1 \cdot a_{33} + 0 \cdot a_{43} \\ 0 \cdot a_{11} + k \cdot a_{21} + 0 \cdot a_{31} + 1 \cdot a_{41} & 0 \cdot a_{12} + k \cdot a_{22} + 0 \cdot a_{32} + 1 \cdot a_{42} & 0 \cdot a_{13} + k \cdot a_{23} + 0 \cdot a_{33} + 1 \cdot a_{43} \end{pmatrix}$$

$$= \begin{pmatrix} a_{11} & a_{12} & a_{13} \\ a_{21} & a_{22} & a_{23} \\ a_{31} & a_{32} & a_{33} \\ a_{41} + ka_{21} & a_{42} + ka_{22} & a_{43} + ka_{23} \end{pmatrix}$$

- Example 2 (k times column 3 is added to column 1.)

$$\begin{pmatrix} a_{11} & a_{12} & a_{13} \\ a_{21} & a_{22} & a_{23} \\ a_{31} & a_{32} & a_{33} \\ a_{41} & a_{42} & a_{43} \end{pmatrix} \begin{pmatrix} 1 & 0 & 0 \\ 0 & 1 & 0 \\ k & 0 & 1 \end{pmatrix}$$

$$= \begin{pmatrix} a_{11}\cdot 1 + a_{12}\cdot 0 + a_{13}\cdot k & a_{11}\cdot 0 + a_{12}\cdot 1 + a_{13}\cdot 0 & a_{11}\cdot 0 + a_{12}\cdot 0 + a_{13}\cdot 1 \\ a_{21}\cdot 1 + a_{22}\cdot 0 + a_{23}\cdot k & a_{21}\cdot 0 + a_{22}\cdot 1 + a_{23}\cdot 0 & a_{21}\cdot 0 + a_{22}\cdot 0 + a_{23}\cdot 1 \\ a_{31}\cdot 1 + a_{32}\cdot 0 + a_{33}\cdot k & a_{31}\cdot 0 + a_{32}\cdot 1 + a_{33}\cdot 0 & a_{31}\cdot 0 + a_{32}\cdot 0 + a_{33}\cdot 1 \\ a_{41}\cdot 1 + a_{42}\cdot 0 + a_{43}\cdot k & a_{41}\cdot 0 + a_{42}\cdot 1 + a_{43}\cdot 0 & a_{41}\cdot 0 + a_{42}\cdot 0 + a_{43}\cdot 1 \end{pmatrix}$$

$$= \begin{pmatrix} a_{11} + ka_{13} & a_{12} & a_{13} \\ a_{21} + ka_{23} & a_{22} & a_{23} \\ a_{31} + ka_{33} & a_{32} & a_{33} \\ a_{41} + ka_{43} & a_{42} & a_{43} \end{pmatrix}$$

FACT 4

The following three $m \times n$ matrices all have the same rank:

1. **The matrix:**

$$\begin{pmatrix} a_{11} & a_{12} & \cdots & a_{1n} \\ a_{21} & a_{22} & \cdots & a_{2n} \\ \vdots & \vdots & \ddots & \vdots \\ a_{m1} & a_{m2} & \cdots & a_{mn} \end{pmatrix}$$

2. **The left product using an invertible $m \times m$ matrix:**

$$\begin{pmatrix} b_{11} & b_{12} & \cdots & b_{1m} \\ b_{21} & b_{22} & \cdots & b_{2m} \\ \vdots & \vdots & \ddots & \vdots \\ b_{m1} & b_{m2} & \cdots & b_{mm} \end{pmatrix} \begin{pmatrix} a_{11} & a_{12} & \cdots & a_{1n} \\ a_{21} & a_{22} & \cdots & a_{2n} \\ \vdots & \vdots & \ddots & \vdots \\ a_{m1} & a_{m2} & \cdots & a_{mn} \end{pmatrix}$$

3. **The right product using an invertible $n \times n$ matrix:**

$$\begin{pmatrix} a_{11} & a_{12} & \cdots & a_{1n} \\ a_{21} & a_{22} & \cdots & a_{2n} \\ \vdots & \vdots & \ddots & \vdots \\ a_{m1} & a_{m2} & \cdots & a_{mn} \end{pmatrix} \begin{pmatrix} c_{11} & c_{12} & \cdots & c_{1n} \\ c_{21} & c_{22} & \cdots & c_{2n} \\ \vdots & \vdots & \ddots & \vdots \\ c_{n1} & c_{n2} & \cdots & c_{nn} \end{pmatrix}$$

In other words, multiplying A by any elementary matrix—on either side—will not change A's rank, since elementary matrices are invertible.

SOLUTION

The following table depicts calculating the rank of the 2×4 matrix:

$$\begin{pmatrix} 1 & 0 & 3 & 1 \\ 0 & 1 & 1 & 2 \end{pmatrix}$$

Begin with $$\begin{pmatrix} 1 & 0 & 3 & 1 \\ 0 & 1 & 1 & 2 \end{pmatrix}$$
Add (–1 · column 2) to column 3 $$\begin{pmatrix} 1 & 0 & 3 & 1 \\ 0 & 1 & 1 & 2 \end{pmatrix}\begin{pmatrix} 1 & 0 & 0 & 0 \\ 0 & 1 & -1 & 0 \\ 0 & 0 & 1 & 0 \\ 0 & 0 & 0 & 1 \end{pmatrix} = \begin{pmatrix} 1 & 0 & 3 & 1 \\ 0 & 1 & 0 & 2 \end{pmatrix}$$
Add (–1 · column 1) to column 4 $$\begin{pmatrix} 1 & 0 & 3 & 1 \\ 0 & 1 & 0 & 2 \end{pmatrix}\begin{pmatrix} 1 & 0 & 0 & -1 \\ 0 & 1 & 0 & 0 \\ 0 & 0 & 1 & 0 \\ 0 & 0 & 0 & 1 \end{pmatrix} = \begin{pmatrix} 1 & 0 & 3 & 0 \\ 0 & 1 & 0 & 2 \end{pmatrix}$$
Add (–3 · column 1) to column 3 $$\begin{pmatrix} 1 & 0 & 3 & 0 \\ 0 & 1 & 0 & 2 \end{pmatrix}\begin{pmatrix} 1 & 0 & -3 & 0 \\ 0 & 1 & 0 & 0 \\ 0 & 0 & 1 & 0 \\ 0 & 0 & 0 & 1 \end{pmatrix} = \begin{pmatrix} 1 & 0 & 0 & 0 \\ 0 & 1 & 0 & 2 \end{pmatrix}$$
Add (–2 · column 2) to column 4 $$\begin{pmatrix} 1 & 0 & 0 & 0 \\ 0 & 1 & 0 & 2 \end{pmatrix}\begin{pmatrix} 1 & 0 & 0 & 0 \\ 0 & 1 & 0 & -2 \\ 0 & 0 & 1 & 0 \\ 0 & 0 & 0 & 1 \end{pmatrix} = \begin{pmatrix} 1 & 0 & 0 & 0 \\ 0 & 1 & 0 & 0 \end{pmatrix}$$

Because of Fact 4, we know that both $\begin{pmatrix} 1 & 0 & 3 & 1 \\ 0 & 1 & 1 & 2 \end{pmatrix}$ and $\begin{pmatrix} 1 & 0 & 0 & 0 \\ 0 & 1 & 0 & 0 \end{pmatrix}$ have the same rank.

One look at the simplified matrix is enough to see that only $\begin{pmatrix} 1 \\ 0 \end{pmatrix}$ and $\begin{pmatrix} 0 \\ 1 \end{pmatrix}$ are linearly independent among its columns.

This means it has a rank of 2, and so does our initial matrix.

THE RELATIONSHIP BETWEEN LINEAR TRANSFORMATIONS AND MATRICES

We talked a bit about the relationship between linear transformations and matrices on page 168. We said that a linear transformation from R^n to R^m could be written as an $m \times n$ matrix:

$$\begin{pmatrix} a_{11} & a_{12} & \cdots & a_{1n} \\ a_{21} & a_{22} & \cdots & a_{2n} \\ \vdots & \vdots & \ddots & \vdots \\ a_{m1} & a_{m2} & \cdots & a_{mn} \end{pmatrix}$$

As you probably noticed, this explanation is a bit vague. The more exact relationship is as follows:

THE RELATIONSHIP BETWEEN LINEAR TRANSFORMATIONS AND MATRICES

If $\begin{pmatrix} x_1 \\ x_2 \\ \vdots \\ x_n \end{pmatrix}$ is an arbitrary element in R^n and f is a function from R^n to R^m,

then f is a linear transformation from R^n to R^m if and only if

$$f\left(\begin{pmatrix} x_1 \\ x_2 \\ \vdots \\ x_n \end{pmatrix}\right) = \begin{pmatrix} a_{11} & a_{12} & \cdots & a_{1n} \\ a_{21} & a_{22} & \cdots & a_{2n} \\ \vdots & \vdots & \ddots & \vdots \\ a_{m1} & a_{m2} & \cdots & a_{mn} \end{pmatrix}\begin{pmatrix} x_1 \\ x_2 \\ \vdots \\ x_n \end{pmatrix}$$

for some matrix A.

8
EIGENVALUES AND EIGENVECTORS

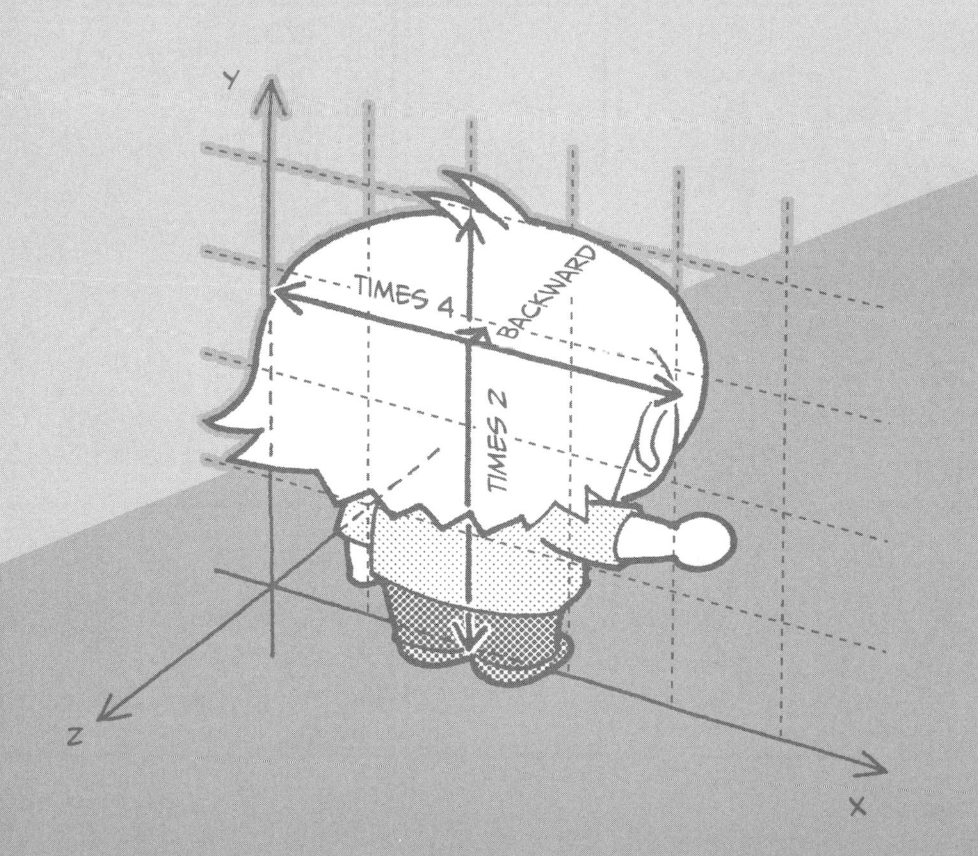

YURINO FROM HANAMICHI UNIVERSITY!
AND JUMONJI FROM NANHOU UNIVERSITY!
南凰
HE LOOKS TOUGH.
STARE
READY!
南凰
I'LL HAVE TO GIVE IT MY ALL.
BEGIN!

BAFF
SEYAAAAAA
SMACK
SHUDDER
GUH...
NGH...
POW
SNUH
I HAVE TO WIN THIS!
I'VE GOT TO SHOW THEM HOW STRONG I CAN BE!
YAAAAAA
ENOUGH!

NANHOU UNIVERSITY!
THANK YOU... VERY MUCH...
DAMN...
GOOD MATCH.

I'M SORRY ABOUT THE MATCH...
YEAH...
MY BROTHER SAID YOU FOUGHT WELL, THOUGH.
REALLY?
DON'T WORRY ABOUT IT.
YOU'LL DO BETTER NEXT TIME.
I KNOW IT!
I'M SORRY! YOU'RE COMPLETELY RIGHT!
SULKING WON'T ACCOMPLISH ANYTHING.

ANYWAY... TODAY'S OUR LAST LESSON.
AND I THOUGHT WE'D WORK ON EIGENVALUES AND EIGENVECTORS.
OKAY. I'M READY FOR ANYTHING!

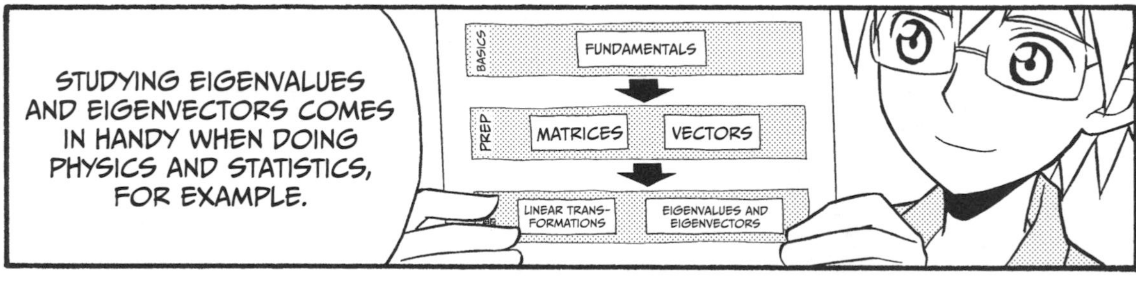
STUDYING EIGENVALUES AND EIGENVECTORS COMES IN HANDY WHEN DOING PHYSICS AND STATISTICS, FOR EXAMPLE.
BASICS
FUNDAMENTALS
PREP
MATRICES
VECTORS
LINEAR TRANS-FORMATIONS
EIGENVALUES AND EIGENVECTORS

THEY ALSO MAKE THESE KINDS OF PROBLEMS MUCH EASIER.
$\begin{pmatrix} a_{11} & a_{12} & \cdots & a_{1n} \\ a_{21} & a_{22} & \cdots & a_{2n} \\ \vdots & \vdots & \ddots & \vdots \\ a_{n1} & a_{n2} & \cdots & a_{nn} \end{pmatrix}^p$
FINDING THE pth POWER OF AN $n \times n$ MATRIX.
IT'S A PRETTY ABSTRACT TOPIC, BUT I'LL TRY TO BE AS CONCRETE AS I CAN.
I APPRECIATE IT!

WHAT ARE EIGENVALUES AND EIGENVECTORS?

OKAY, FIRST PROBLEM. FIND THE IMAGE OF

$$c_1 \begin{pmatrix} 3 \\ 1 \end{pmatrix} + c_2 \begin{pmatrix} 1 \\ 2 \end{pmatrix}$$

USING THE LINEAR TRANSFORMATION DETERMINED BY THE 2×2 MATRIX

$$\begin{pmatrix} 8 & -3 \\ 2 & 1 \end{pmatrix}$$

(WHERE c_1 AND c_2 ARE REAL NUMBERS).

$$\begin{pmatrix} 8 & -3 \\ 2 & 1 \end{pmatrix} \left[c_1 \begin{pmatrix} 3 \\ 1 \end{pmatrix} + c_2 \begin{pmatrix} 1 \\ 2 \end{pmatrix} \right]$$

$$= c_1 \begin{pmatrix} 8 & -3 \\ 2 & 1 \end{pmatrix} \begin{pmatrix} 3 \\ 1 \end{pmatrix} + c_2 \begin{pmatrix} 8 & -3 \\ 2 & 1 \end{pmatrix} \begin{pmatrix} 1 \\ 2 \end{pmatrix}$$

$$= c_1 \begin{pmatrix} 8 \cdot 3 + (-3) \cdot 1 \\ 2 \cdot 3 + \quad 1 \cdot 1 \end{pmatrix} + c_2 \begin{pmatrix} 8 \cdot 1 + (-3) \cdot 2 \\ 2 \cdot 1 + \quad 1 \cdot 2 \end{pmatrix}$$

$$= c_1 \begin{pmatrix} 21 \\ 7 \end{pmatrix} + c_2 \begin{pmatrix} 2 \\ 4 \end{pmatrix}$$

LIKE THIS?

SO CLOSE!

UH, LIKE THIS?

$$= c_1 \begin{pmatrix} 21 \\ 7 \end{pmatrix} + c_2 \begin{pmatrix} 2 \\ 4 \end{pmatrix}$$

$$= c_1 \left[7 \begin{pmatrix} 3 \\ 1 \end{pmatrix} \right] + c_2 \left[2 \begin{pmatrix} 1 \\ 2 \end{pmatrix} \right]$$

EXACTLY!

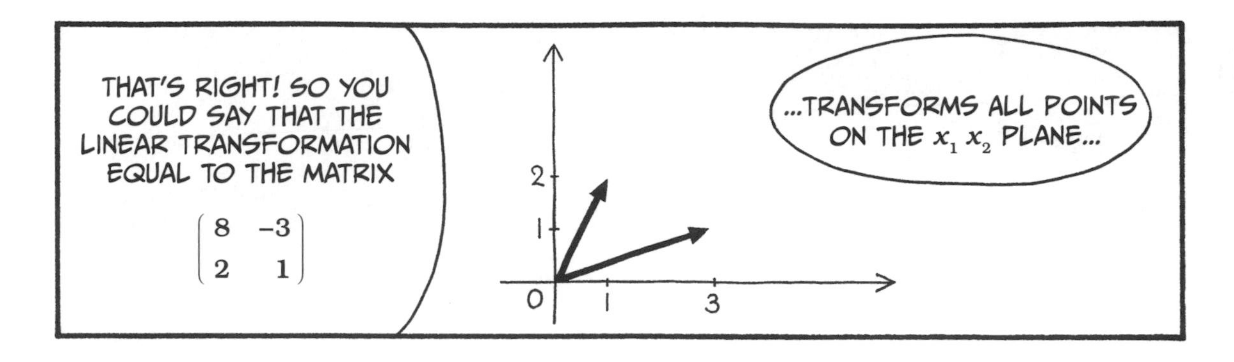
THAT'S RIGHT! SO YOU COULD SAY THAT THE LINEAR TRANSFORMATION EQUAL TO THE MATRIX
$\begin{pmatrix} 8 & -3 \\ 2 & 1 \end{pmatrix}$
...TRANSFORMS ALL POINTS ON THE $x_1 x_2$ PLANE...
2
1
0
1
3

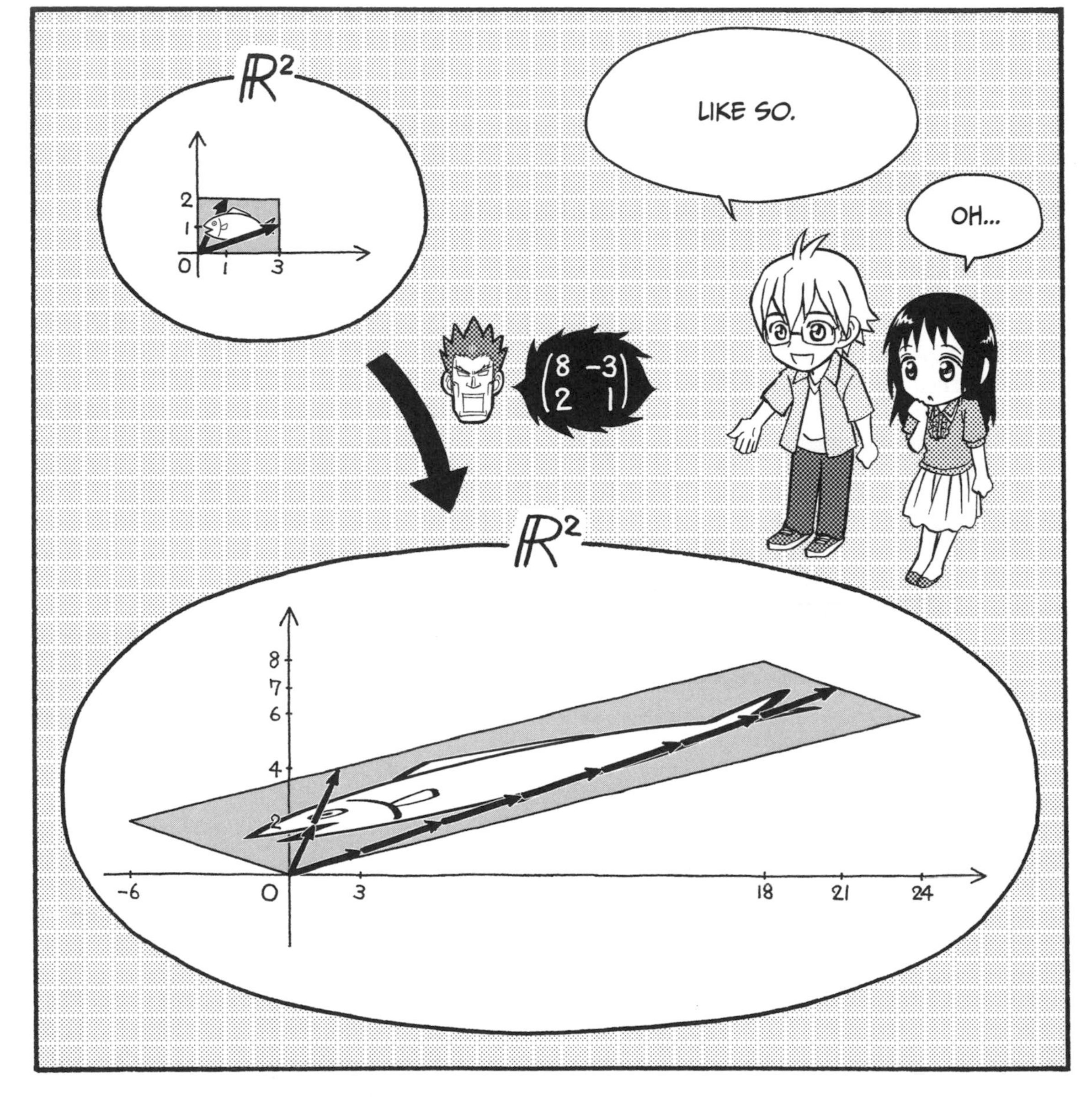
$\mathbb{R}^2$
2
1
0
1
3
LIKE SO.
OH...
$\begin{pmatrix} 8 & -3 \\ 2 & 1 \end{pmatrix}$
$\mathbb{R}^2$
8
7
6
4
2
-6
O
3
18
21
24

LET'S MOVE ON TO ANOTHER PROBLEM.

FIND THE IMAGE OF $c_1\begin{pmatrix}1\\0\\0\end{pmatrix}+c_2\begin{pmatrix}0\\1\\0\end{pmatrix}+c_3\begin{pmatrix}0\\0\\1\end{pmatrix}$ USING

THE LINEAR TRANSFORMATION DETERMINED BY THE 3×3 MATRIX $\begin{pmatrix}4&0&0\\0&2&0\\0&0&-1\end{pmatrix}$

(WHERE c_1, c_2, AND c_3 ARE REAL NUMBERS).

$$\begin{pmatrix}4&0&0\\0&2&0\\0&0&-1\end{pmatrix}\left[c_1\begin{pmatrix}1\\0\\0\end{pmatrix}+c_2\begin{pmatrix}0\\1\\0\end{pmatrix}+c_3\begin{pmatrix}0\\0\\1\end{pmatrix}\right]$$

$$=c_1\begin{pmatrix}4&0&0\\0&2&0\\0&0&-1\end{pmatrix}\begin{pmatrix}1\\0\\0\end{pmatrix}+c_2\begin{pmatrix}4&0&0\\0&2&0\\0&0&-1\end{pmatrix}\begin{pmatrix}0\\1\\0\end{pmatrix}+c_3\begin{pmatrix}4&0&0\\0&2&0\\0&0&-1\end{pmatrix}\begin{pmatrix}0\\0\\1\end{pmatrix}$$

$$=c_1\begin{pmatrix}4\\0\\0\end{pmatrix}+c_2\begin{pmatrix}0\\2\\0\end{pmatrix}+c_3\begin{pmatrix}0\\0\\-1\end{pmatrix}$$

$$=c_1\left[4\begin{pmatrix}1\\0\\0\end{pmatrix}\right]+c_2\left[2\begin{pmatrix}0\\1\\0\end{pmatrix}\right]+c_3\left[-\begin{pmatrix}0\\0\\1\end{pmatrix}\right]$$

LIKE THIS?

CORRECT.

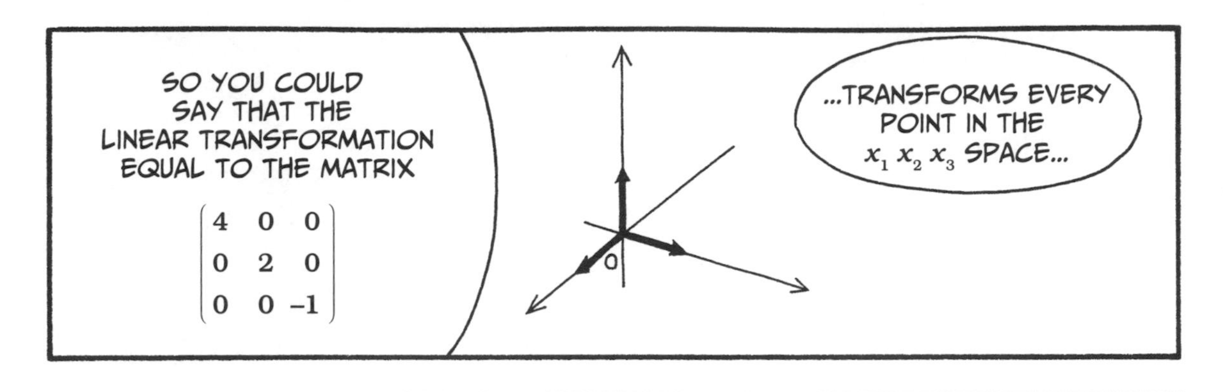
SO YOU COULD SAY THAT THE LINEAR TRANSFORMATION EQUAL TO THE MATRIX
$\begin{pmatrix} 4 & 0 & 0 \\ 0 & 2 & 0 \\ 0 & 0 & -1 \end{pmatrix}$
...TRANSFORMS EVERY POINT IN THE $x_1\ x_2\ x_3$ SPACE...
0

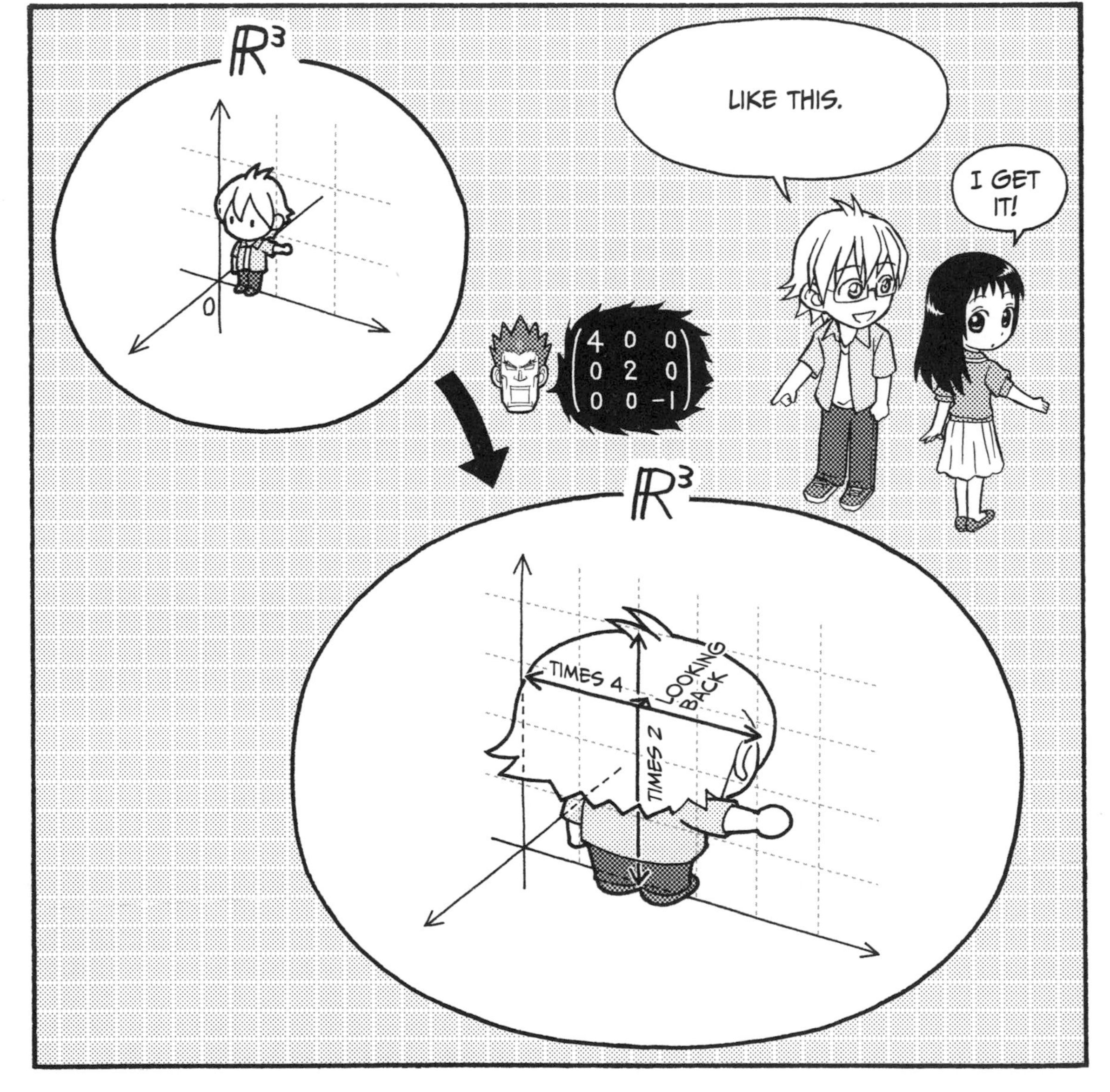
$\mathbb{R}^3$
0
LIKE THIS.
I GET IT!
$\begin{pmatrix} 4 & 0 & 0 \\ 0 & 2 & 0 \\ 0 & 0 & -1 \end{pmatrix}$
$\mathbb{R}^3$
TIMES 4
LOOKING BACK
TIMES 2

EIGENVALUES AND EIGENVECTORS

If the image of a vector $\begin{pmatrix} x_1 \\ x_2 \\ \vdots \\ x_n \end{pmatrix}$ through the linear transformation determined by the matrix

$\begin{pmatrix} a_{11} & a_{12} & \cdots & a_{1n} \\ a_{21} & a_{22} & \cdots & a_{2n} \\ \vdots & \vdots & \ddots & \vdots \\ a_{n1} & a_{n2} & \cdots & a_{nn} \end{pmatrix}$ is equal to $\lambda \begin{pmatrix} x_1 \\ x_2 \\ \vdots \\ x_n \end{pmatrix}$, λ is said to be an *eigenvalue* to the matrix,

and $\begin{pmatrix} x_1 \\ x_2 \\ \vdots \\ x_n \end{pmatrix}$ is said to be an *eigenvector* corresponding to the eigenvalue λ.

The zero vector can never be an eigenvector.

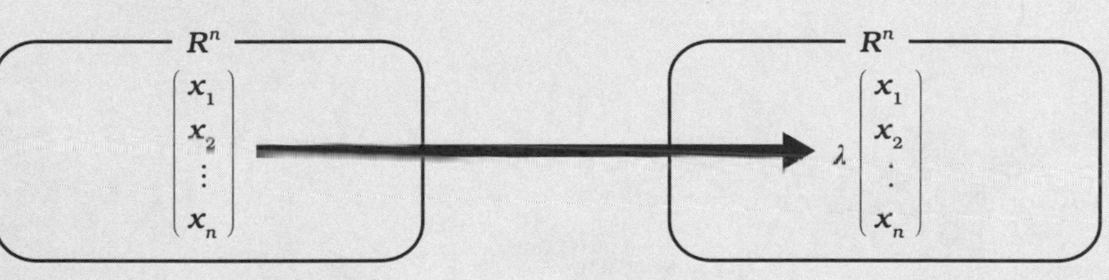

SO THE TWO EXAMPLES COULD BE SUMMARIZED LIKE THIS?

EXACTLY!

MATRIX	$\begin{pmatrix} 8 & -3 \\ 2 & 1 \end{pmatrix}$	$\begin{pmatrix} 4 & 0 & 0 \\ 0 & 2 & 0 \\ 0 & 0 & -1 \end{pmatrix}$
EIGENVALUE	$\lambda = 7, 2$	$\lambda = 4, 2, -1$
EIGENVECTOR	THE VECTOR CORRESPONDING TO $\lambda = 7$ $\begin{pmatrix} 3 \\ 1 \end{pmatrix}$ THE VECTOR CORRESPONDING TO $\lambda = 2$ $\begin{pmatrix} 1 \\ 2 \end{pmatrix}$	THE VECTOR CORRESPONDING TO $\lambda = 4$ $\begin{pmatrix} 1 \\ 0 \\ 0 \end{pmatrix}$ THE VECTOR CORRESPONDING TO $\lambda = 2$ $\begin{pmatrix} 0 \\ 1 \\ 0 \end{pmatrix}$ THE VECTOR CORRESPONDING TO $\lambda = 1$ $\begin{pmatrix} 0 \\ 0 \\ 1 \end{pmatrix}$

CALCULATING EIGENVALUES AND EIGENVECTORS

THE RELATIONSHIP BETWEEN THE DETERMINANT AND EIGENVALUES OF A MATRIX

λ is an eigenvalue of the matrix

$$\begin{pmatrix} a_{11} & a_{12} & \cdots & a_{1n} \\ a_{21} & a_{22} & \cdots & a_{2n} \\ \vdots & \vdots & \ddots & \vdots \\ a_{n1} & a_{n2} & \cdots & a_{nn} \end{pmatrix} \text{ if and only if } \det \begin{pmatrix} a_{11}-\lambda & a_{12} & \cdots & a_{1n} \\ a_{21} & a_{22}-\lambda & \cdots & a_{2n} \\ \vdots & \vdots & \ddots & \vdots \\ a_{n1} & a_{n2} & \cdots & a_{nn}-\lambda \end{pmatrix} = 0$$

$$\det\begin{pmatrix} 8-\lambda & -3 \\ 2 & 1-\lambda \end{pmatrix} = (8-\lambda)\cdot(1-\lambda)-(-3)\cdot 2$$

$$= (\lambda-8)\cdot(\lambda-1)-(-3)\cdot 2$$

$$= \lambda^2 - 9\lambda + 8 + 6$$

$$= \lambda^2 - 9\lambda + 14$$

$$= (\lambda-7)(\lambda-2) = 0$$

$$\lambda = 7, 2$$

SO...

$$\begin{pmatrix} 8 & -3 \\ 2 & 1 \end{pmatrix}\begin{pmatrix} x_1 \\ x_2 \end{pmatrix} = \lambda\begin{pmatrix} x_1 \\ x_2 \end{pmatrix}, \text{ THAT IS } \begin{pmatrix} 8-\lambda & -3 \\ 2 & 1-\lambda \end{pmatrix}\begin{pmatrix} x_1 \\ x_2 \end{pmatrix} = \begin{pmatrix} 0 \\ 0 \end{pmatrix}$$

PROBLEM 1

Find an eigenvector corresponding to $\lambda = 7$.

Let's plug our value into the formula:

$$\begin{pmatrix} 8-7 & -3 \\ 2 & 1-7 \end{pmatrix}\begin{pmatrix} x_1 \\ x_2 \end{pmatrix} = \begin{pmatrix} 1 & -3 \\ 2 & -6 \end{pmatrix}\begin{pmatrix} x_1 \\ x_2 \end{pmatrix} = \begin{pmatrix} x_1 - 3x_2 \\ 2x_1 - 6x_2 \end{pmatrix} = [x_1 - 3x_2]\begin{pmatrix} 1 \\ 2 \end{pmatrix} = \begin{pmatrix} 0 \\ 0 \end{pmatrix}$$

This means that $x_1 = 3x_2$, which leads us to our eigenvector

$$\begin{pmatrix} x_1 \\ x_2 \end{pmatrix} = \begin{pmatrix} 3c_1 \\ c_1 \end{pmatrix} = c_1\begin{pmatrix} 3 \\ 1 \end{pmatrix}$$

where c_1 is an arbitrary nonzero real number.

PROBLEM 2

Find an eigenvector corresponding to $\lambda = 2$.

Let's plug our value into the formula:

$$\begin{pmatrix} 8-2 & -3 \\ 2 & 1-2 \end{pmatrix}\begin{pmatrix} x_1 \\ x_2 \end{pmatrix} = \begin{pmatrix} 6 & -3 \\ 2 & -1 \end{pmatrix}\begin{pmatrix} x_1 \\ x_2 \end{pmatrix} = \begin{pmatrix} 6x_1 - 3x_2 \\ 2x_1 - x_2 \end{pmatrix} = [2x_1 - x_2]\begin{pmatrix} 3 \\ 1 \end{pmatrix} = \begin{pmatrix} 0 \\ 0 \end{pmatrix}$$

This means that $x_2 = 2x_1$, which leads us to our eigenvector

$$\begin{pmatrix} x_1 \\ x_2 \end{pmatrix} = \begin{pmatrix} c_2 \\ 2c_2 \end{pmatrix} = c_2\begin{pmatrix} 1 \\ 2 \end{pmatrix}$$

where c_2 is an arbitrary nonzero real number.

CALCULATING THE PTH POWER OF AN NxN MATRIX

IT'S FINALLY TIME TO TACKLE TODAY'S REAL PROBLEM! FINDING THE pth POWER OF AN $n \times n$ MATRIX.

$$\begin{pmatrix} a_{11} & a_{12} & \cdots & a_{1n} \\ a_{21} & a_{22} & \cdots & a_{2n} \\ \vdots & \vdots & \ddots & \vdots \\ a_{n1} & a_{n2} & \cdots & a_{nn} \end{pmatrix}^p$$

$$\begin{pmatrix} 8 & -3 \\ 2 & 1 \end{pmatrix}$$

$$\begin{pmatrix} 8 & -3 \\ 2 & 1 \end{pmatrix}\begin{pmatrix} x_1 \\ x_2 \end{pmatrix} = \lambda \begin{pmatrix} x_1 \\ x_2 \end{pmatrix}$$

$$\begin{pmatrix} 8 & -3 \\ 2 & 1 \end{pmatrix}\begin{pmatrix} 3 \\ 1 \end{pmatrix} = 7\begin{pmatrix} 3 \\ 1 \end{pmatrix} = \begin{pmatrix} 3 \cdot 7 \\ 1 \cdot 7 \end{pmatrix} \qquad \begin{pmatrix} 8 & -3 \\ 2 & 1 \end{pmatrix}\begin{pmatrix} 1 \\ 2 \end{pmatrix} = 2\begin{pmatrix} 1 \\ 2 \end{pmatrix} = \begin{pmatrix} 1 \cdot 2 \\ 2 \cdot 2 \end{pmatrix}$$

FOR SIMPLICITY'S SAKE, LET'S CHOOSE $c_1 = c_2 = 1$.

$$\begin{pmatrix} 8 & -3 \\ 2 & 1 \end{pmatrix}\begin{pmatrix} 3 & 1 \\ 1 & 2 \end{pmatrix} = \begin{pmatrix} 3 \cdot 7 & 1 \cdot 2 \\ 1 \cdot 7 & 2 \cdot 2 \end{pmatrix}$$

USING THE TWO CALCULATIONS ABOVE...

$$= \begin{pmatrix} 3 & 1 \\ 1 & 2 \end{pmatrix}\begin{pmatrix} 7 & 0 \\ 0 & 2 \end{pmatrix}$$

LET'S MULTIPLY $\begin{pmatrix} 3 & 1 \\ 1 & 2 \end{pmatrix}^{-1}$ TO THE RIGHT OF BOTH SIDES OF THE EQUATION. REFER TO PAGE 91 TO SEE WHY $\begin{pmatrix} 3 & 1 \\ 1 & 2 \end{pmatrix}^{-1}$ EXISTS.

$$\begin{pmatrix} 8 & -3 \\ 2 & 1 \end{pmatrix}\begin{pmatrix} 3 & 1 \\ 1 & 2 \end{pmatrix}\begin{pmatrix} 3 & 1 \\ 1 & 2 \end{pmatrix}^{-1} = \begin{pmatrix} 3 & 1 \\ 1 & 2 \end{pmatrix}\begin{pmatrix} 7 & 0 \\ 0 & 2 \end{pmatrix}\begin{pmatrix} 3 & 1 \\ 1 & 2 \end{pmatrix}^{-1}$$

$$\begin{pmatrix} 8 & -3 \\ 2 & 1 \end{pmatrix} = \begin{pmatrix} 3 & 1 \\ 1 & 2 \end{pmatrix}\begin{pmatrix} 7 & 0 \\ 0 & 2 \end{pmatrix}\begin{pmatrix} 3 & 1 \\ 1 & 2 \end{pmatrix}^{-1}$$

TRY USING THE FORMULA TO CALCULATE
$\begin{pmatrix} 8 & -3 \\ 2 & 1 \end{pmatrix}^2$
HMM... OKAY.

$\begin{pmatrix} 8 & -3 \\ 2 & 1 \end{pmatrix}^2$
$= \begin{pmatrix} 8 & -3 \\ 2 & 1 \end{pmatrix}\begin{pmatrix} 8 & -3 \\ 2 & 1 \end{pmatrix}$
$= \begin{pmatrix} 3 & 1 \\ 1 & 2 \end{pmatrix}\begin{pmatrix} 7 & 0 \\ 0 & 2 \end{pmatrix}\begin{pmatrix} 3 & 1 \\ 1 & 2 \end{pmatrix}^{-1}\begin{pmatrix} 3 & 1 \\ 1 & 2 \end{pmatrix}\begin{pmatrix} 7 & 0 \\ 0 & 2 \end{pmatrix}\begin{pmatrix} 3 & 1 \\ 1 & 2 \end{pmatrix}^{-1}$
$= \begin{pmatrix} 3 & 1 \\ 1 & 2 \end{pmatrix}\begin{pmatrix} 7 & 0 \\ 0 & 2 \end{pmatrix}^2\begin{pmatrix} 3 & 1 \\ 1 & 2 \end{pmatrix}^{-1}$
$= \begin{pmatrix} 3 & 1 \\ 1 & 2 \end{pmatrix}\begin{pmatrix} 7^2 & 0 \\ 0 & 2^2 \end{pmatrix}\begin{pmatrix} 3 & 1 \\ 1 & 2 \end{pmatrix}^{-1}$
IS...THIS IT?
YEP!
YAY!

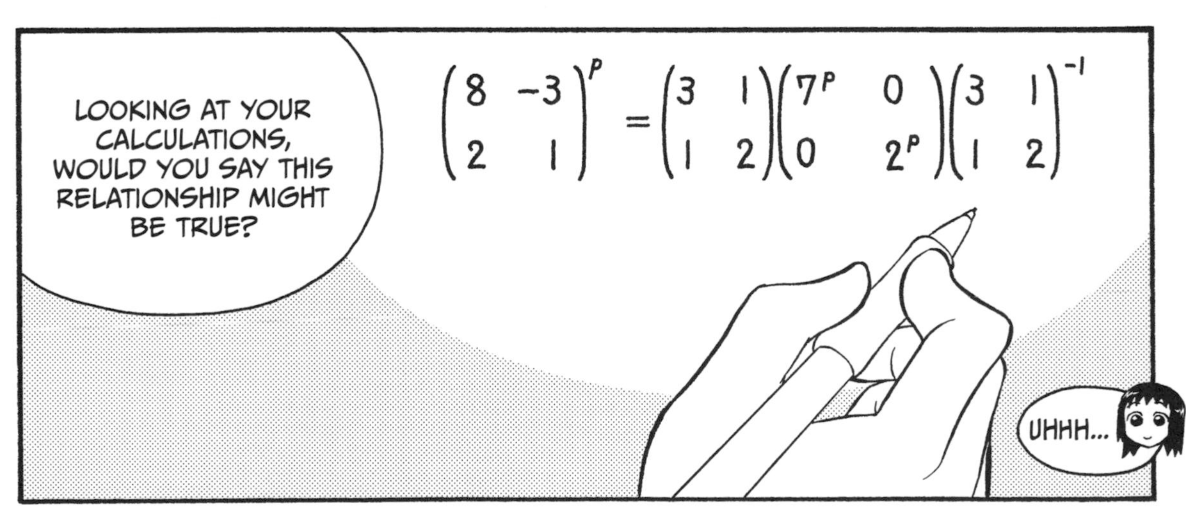
LOOKING AT YOUR CALCULATIONS, WOULD YOU SAY THIS RELATIONSHIP MIGHT BE TRUE?
$\begin{pmatrix} 8 & -3 \\ 2 & 1 \end{pmatrix}^p = \begin{pmatrix} 3 & 1 \\ 1 & 2 \end{pmatrix}\begin{pmatrix} 7^p & 0 \\ 0 & 2^p \end{pmatrix}\begin{pmatrix} 3 & 1 \\ 1 & 2 \end{pmatrix}^{-1}$
UHHH...

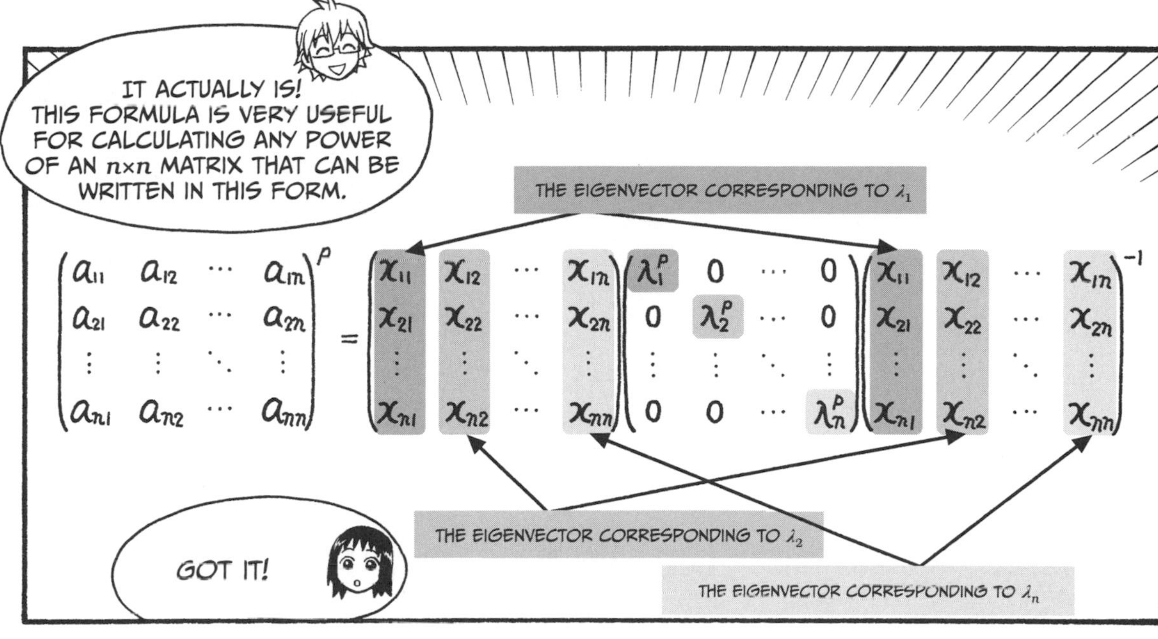

IT ACTUALLY IS! THIS FORMULA IS VERY USEFUL FOR CALCULATING ANY POWER OF AN $n \times n$ MATRIX THAT CAN BE WRITTEN IN THIS FORM.
THE EIGENVECTOR CORRESPONDING TO λ_1
$$\begin{pmatrix} a_{11} & a_{12} & \cdots & a_{1n} \\ a_{21} & a_{22} & \cdots & a_{2n} \\ \vdots & \vdots & \ddots & \vdots \\ a_{n1} & a_{n2} & \cdots & a_{nn} \end{pmatrix}^p = \begin{pmatrix} x_{11} & x_{12} & \cdots & x_{1n} \\ x_{21} & x_{22} & \cdots & x_{2n} \\ \vdots & \vdots & \ddots & \vdots \\ x_{n1} & x_{n2} & \cdots & x_{nn} \end{pmatrix} \begin{pmatrix} \lambda_1^p & 0 & \cdots & 0 \\ 0 & \lambda_2^p & \cdots & 0 \\ \vdots & \vdots & \ddots & \vdots \\ 0 & 0 & \cdots & \lambda_n^p \end{pmatrix} \begin{pmatrix} x_{11} & x_{12} & \cdots & x_{1n} \\ x_{21} & x_{22} & \cdots & x_{2n} \\ \vdots & \vdots & \ddots & \vdots \\ x_{n1} & x_{n2} & \cdots & x_{nn} \end{pmatrix}^{-1}$$
THE EIGENVECTOR CORRESPONDING TO λ_2
THE EIGENVECTOR CORRESPONDING TO λ_n
GOT IT!

OH, AND BY THE WAY...

WHEN $p = 1$, WE SAY THAT THE FORMULA *DIAGONALIZES* THE $n \times n$ MATRIX
$$\begin{pmatrix} a_{11} & a_{12} & \cdots & a_{1n} \\ a_{21} & a_{22} & \cdots & a_{2n} \\ \vdots & \vdots & \ddots & \vdots \\ a_{n1} & a_{n2} & \cdots & a_{nn} \end{pmatrix}$$
AND THAT'S IT!
$$\begin{pmatrix} x_{11} & x_{12} & \cdots & x_{1n} \\ x_{21} & x_{22} & \cdots & x_{2n} \\ \vdots & \vdots & \ddots & \vdots \\ x_{n1} & x_{n2} & \cdots & x_{nn} \end{pmatrix}^{-1} \begin{pmatrix} a_{11} & a_{12} & \cdots & a_{1n} \\ a_{21} & a_{22} & \cdots & a_{2n} \\ \vdots & \vdots & \ddots & \vdots \\ a_{n1} & a_{n2} & \cdots & a_{nn} \end{pmatrix} \begin{pmatrix} x_{11} & x_{12} & \cdots & x_{1n} \\ x_{21} & x_{22} & \cdots & x_{2n} \\ \vdots & \vdots & \ddots & \vdots \\ x_{n1} & x_{n2} & \cdots & x_{nn} \end{pmatrix} = \begin{pmatrix} \lambda_1 & 0 & \cdots & 0 \\ 0 & \lambda_2 & \cdots & 0 \\ \vdots & \vdots & \ddots & \vdots \\ 0 & 0 & \cdots & \lambda_n \end{pmatrix}$$
THE RIGHT SIDE OF THE EQUATION IS THE *DIAGONALIZED* FORM OF THE MIDDLE MATRIX ON THE LEFT SIDE.
NICE!

THAT WAS THE LAST LESSON!
HOW DO YOU FEEL? DID YOU GET THE GIST OF IT?

YEAH, THANKS TO YOU.
AWESOME!

REALLY, THOUGH, THANKS FOR HELPING ME OUT.
I KNOW YOU'RE BUSY, AND YOU'VE BEEN AWFULLY TIRED BECAUSE OF YOUR KARATE PRACTICE.

NOT AT ALL! HOW COULD I POSSIBLY HAVE BEEN TIRED AFTER ALL THAT WONDERFUL FOOD YOU GAVE ME?
I SHOULD BE THANKING *YOU*!

I'LL MISS THESE SESSIONS, YOU KNOW! MY AFTERNOONS WILL BE SO LONELY FROM NOW ON...

WELL...WE COULD GO OUT SOMETIME...
HMM?

YEAH...TO LOOK FOR MATH BOOKS, OR SOMETHING... YOU KNOW...
IF YOU DON'T HAVE ANYTHING ELSE TO DO...

SURE, SOUNDS LIKE FUN!

SO WHEN WOULD YOU LIKE TO GO?

MULTIPLICITY AND DIAGONALIZATION

We said on page 221 that any $n \times n$ matrix could be expressed in this form:

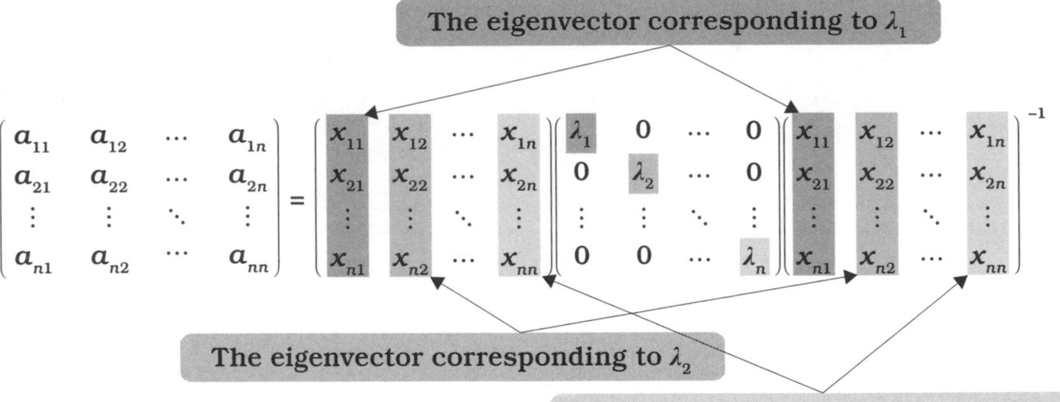

This isn't totally true, as the concept of *multiplicity*[1] plays a large role in whether a matrix can be diagonalized or not. For example, if all n solutions of the following equation

$$\det \begin{pmatrix} a_{11} - \lambda & a_{12} & \cdots & a_{1n} \\ a_{21} & a_{22} - \lambda & \cdots & a_{2n} \\ \vdots & \vdots & \ddots & \vdots \\ a_{n1} & a_{n2} & \cdots & a_{nn} - \lambda \end{pmatrix} = 0$$

are real and have multiplicity 1, then diagonalization is possible. The situation becomes more complicated when we have to deal with eigenvalues that have multiplicity greater than 1. We will therefore look at a few examples involving:

- Matrices with eigenvalues having multiplicity greater than 1 that can be diagonalized
- Matrices with eigenvalues having multiplicity greater than 1 that cannot be diagonalized

1. The multiplicity of any polynomial root reveals how many identical copies of that same root exist in the polynomial. For instance, in the polynomial $f(x) = (x - 1)^4(x + 2)^2x$, the factor $(x - 1)$ has multiplicity 4, $(x + 2)$ has 2, and x has 1.

A DIAGONALIZABLE MATRIX WITH AN EIGENVALUE HAVING MULTIPLICITY 2

PROBLEM

Use the following matrix in both problems:

$$\begin{pmatrix} 1 & 0 & 0 \\ 1 & 1 & -1 \\ -2 & 0 & 3 \end{pmatrix}$$

1. Find all eigenvalues and eigenvectors of the matrix.
2. Express the matrix in the following form:

$$\begin{pmatrix} x_{11} & x_{12} & x_{13} \\ x_{21} & x_{22} & x_{23} \\ x_{31} & x_{32} & x_{33} \end{pmatrix} \begin{pmatrix} \lambda_1 & 0 & 0 \\ 0 & \lambda_2 & 0 \\ 0 & 0 & \lambda_3 \end{pmatrix} \begin{pmatrix} x_{11} & x_{12} & x_{13} \\ x_{21} & x_{22} & x_{23} \\ x_{31} & x_{32} & x_{33} \end{pmatrix}^{-1}$$

SOLUTION

1. The eigenvalues λ of the 3×3 matrix

$$\begin{pmatrix} 1 & 0 & 0 \\ 1 & 1 & -1 \\ 2 & 0 & 3 \end{pmatrix}$$

are the roots of the characteristic equation: $\det \begin{pmatrix} 1-\lambda & 0 & 0 \\ 1 & 1-\lambda & -1 \\ -2 & 0 & 3-\lambda \end{pmatrix} = 0.$

$$\det \begin{pmatrix} 1-\lambda & 0 & 0 \\ 1 & 1-\lambda & -1 \\ -2 & 0 & 3-\lambda \end{pmatrix}$$

$$= (1-\lambda)(1-\lambda)(3-\lambda) + 0 \cdot (-1) \cdot (-2) + 0 \cdot 1 \cdot 0$$
$$- 0 \cdot (1-\lambda) \cdot (-2) - 0 \cdot 1 \cdot (3-\lambda) - (1-\lambda) \cdot (-1) \cdot 0$$
$$= (1-\lambda)^2(3-\lambda) = 0$$

$$\lambda = 3, 1$$

Note that the eigenvalue 1 has multiplicity 2.

A. The eigenvectors corresponding to $\lambda = 3$

Let's insert our eigenvalue into the following formula:

$$\begin{pmatrix} 1 & 0 & 0 \\ 1 & 1 & -1 \\ -2 & 0 & 3 \end{pmatrix}\begin{pmatrix} x_1 \\ x_2 \\ x_3 \end{pmatrix} = \lambda \begin{pmatrix} x_1 \\ x_2 \\ x_3 \end{pmatrix}, \text{ that is } \begin{pmatrix} 1-\lambda & 0 & 0 \\ 1 & 1-\lambda & -1 \\ -2 & 0 & 3-\lambda \end{pmatrix}\begin{pmatrix} x_1 \\ x_2 \\ x_3 \end{pmatrix} = \begin{pmatrix} 0 \\ 0 \\ 0 \end{pmatrix}$$

This gives us:

$$\begin{pmatrix} 1-3 & 0 & 0 \\ 1 & 1-3 & -1 \\ -2 & 0 & 3-3 \end{pmatrix}\begin{pmatrix} x_1 \\ x_2 \\ x_3 \end{pmatrix} = \begin{pmatrix} -2 & 0 & 0 \\ 1 & -2 & -1 \\ -2 & 0 & 0 \end{pmatrix}\begin{pmatrix} x_1 \\ x_2 \\ x_3 \end{pmatrix} = \begin{pmatrix} -2x_1 \\ x_1 - 2x_2 - x_3 \\ -2x_1 \end{pmatrix} = \begin{pmatrix} 0 \\ 0 \\ 0 \end{pmatrix}$$

The solutions are as follows:

$$\begin{cases} x_1 = 0 \\ x_3 = -2x_2 \end{cases} \text{ and the eigenvector } \begin{pmatrix} x_1 \\ x_2 \\ x_3 \end{pmatrix} = \begin{pmatrix} 0 \\ c_1 \\ -2c_1 \end{pmatrix} = c_1 \begin{pmatrix} 0 \\ 1 \\ -2 \end{pmatrix}$$

where c_1 is a real nonzero number.

B. The eigenvectors corresponding to $\lambda = 1$

Repeating the steps above, we get

$$\begin{pmatrix} 1-1 & 0 & 0 \\ 1 & 1-1 & -1 \\ -2 & 0 & 3-1 \end{pmatrix}\begin{pmatrix} x_1 \\ x_2 \\ x_3 \end{pmatrix} = \begin{pmatrix} 0 & 0 & 0 \\ 1 & 0 & -1 \\ -2 & 0 & 2 \end{pmatrix}\begin{pmatrix} x_1 \\ x_2 \\ x_3 \end{pmatrix} = \begin{pmatrix} 0 \\ x_1 - x_3 \\ -2x_1 + 2x_3 \end{pmatrix} = \begin{pmatrix} 0 \\ 0 \\ 0 \end{pmatrix}$$

and see that $x_3 = x_1$ and x_2 can be any real number. The eigenvector consequently becomes

$$\begin{pmatrix} x_1 \\ x_2 \\ x_3 \end{pmatrix} = \begin{pmatrix} c_1 \\ c_2 \\ c_1 \end{pmatrix} = c_1 \begin{pmatrix} 1 \\ 0 \\ 1 \end{pmatrix} + c_2 \begin{pmatrix} 0 \\ 1 \\ 0 \end{pmatrix}$$

where c_1 and c_2 are arbitrary real numbers that cannot both be zero.

3. We then apply the formula from page 221:

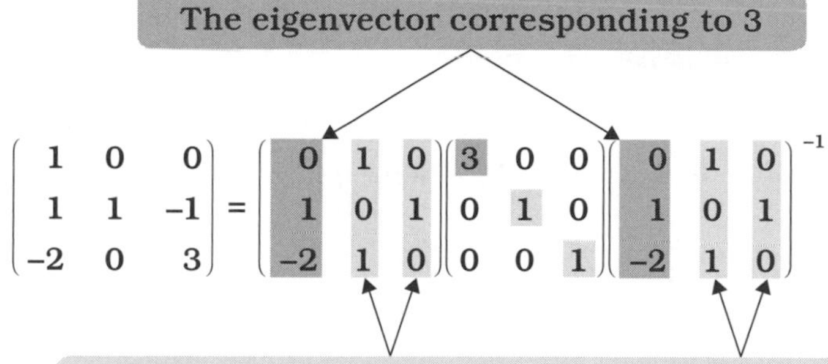

$$\begin{pmatrix} 1 & 0 & 0 \\ 1 & 1 & -1 \\ -2 & 0 & 3 \end{pmatrix} = \begin{pmatrix} 0 & 1 & 0 \\ 1 & 0 & 1 \\ -2 & 1 & 0 \end{pmatrix} \begin{pmatrix} 3 & 0 & 0 \\ 0 & 1 & 0 \\ 0 & 0 & 1 \end{pmatrix} \begin{pmatrix} 0 & 1 & 0 \\ 1 & 0 & 1 \\ -2 & 1 & 0 \end{pmatrix}^{-1}$$

A NON-DIAGONALIZABLE MATRIX WITH A REAL EIGENVALUE HAVING MULTIPLICITY 2

PROBLEM

Use the following matrix in both problems:

$$\begin{pmatrix} 1 & 0 & 0 \\ -7 & 1 & -1 \\ 4 & 0 & 3 \end{pmatrix}$$

1. Find all eigenvalues and eigenvectors of the matrix.
2. Express the matrix in the following form.

$$\begin{pmatrix} x_{11} & x_{12} & x_{13} \\ x_{21} & x_{22} & x_{23} \\ x_{31} & x_{32} & x_{33} \end{pmatrix} \begin{pmatrix} \lambda_1 & 0 & 0 \\ 0 & \lambda_2 & 0 \\ 0 & 0 & \lambda_3 \end{pmatrix} \begin{pmatrix} x_{11} & x_{12} & x_{13} \\ x_{21} & x_{22} & x_{23} \\ x_{31} & x_{32} & x_{33} \end{pmatrix}^{-1}$$

SOLUTION

1. The eigenvalues λ of the 3×3 matrix

$$\begin{pmatrix} 1 & 0 & 0 \\ -7 & 1 & -1 \\ 4 & 0 & 3 \end{pmatrix}$$

are the roots of the characteristic equation: $\det \begin{pmatrix} 1-\lambda & 0 & 0 \\ -7 & 1-\lambda & -1 \\ 4 & 0 & 3-\lambda \end{pmatrix} = 0.$

$$\det\begin{pmatrix}1-\lambda & 0 & 0\\ -7 & 1-\lambda & -1\\ 4 & 0 & 3-\lambda\end{pmatrix}$$

$$\begin{aligned}&= (1-\lambda)(1-\lambda)(3-\lambda) + 0\cdot(-1)\cdot 4 + 0\cdot(-7)\cdot 0\\ &\quad - 0\cdot(1-\lambda)\cdot 4 - 0\cdot(-7)\cdot(3-\lambda) - (1-\lambda)\cdot(-1)\cdot 0\\ &= (1-\lambda)^2(3-\lambda) = 0\end{aligned}$$

$$\lambda = 3,\ 1$$

Again, note that the eigenvalue 1 has multiplicity 2.

A. The eigenvectors corresponding to $\lambda = 3$

Let's insert our eigenvalue into the following formula:

$$\begin{pmatrix}1 & 0 & 0\\ -7 & 1 & -1\\ 4 & 0 & 3\end{pmatrix}\begin{pmatrix}x_1\\ x_2\\ x_3\end{pmatrix} = \lambda\begin{pmatrix}x_1\\ x_2\\ x_3\end{pmatrix}, \text{ that is } \begin{pmatrix}1-\lambda & 0 & 0\\ -7 & 1-\lambda & -1\\ 4 & 0 & 3-\lambda\end{pmatrix}\begin{pmatrix}x_1\\ x_2\\ x_3\end{pmatrix} = \begin{pmatrix}0\\ 0\\ 0\end{pmatrix}$$

This gives us

$$\begin{pmatrix}1-3 & 0 & 0\\ -7 & 1-3 & -1\\ 4 & 0 & 3-3\end{pmatrix}\begin{pmatrix}x_1\\ x_2\\ x_3\end{pmatrix} = \begin{pmatrix}-2 & 0 & 0\\ -7 & -2 & -1\\ 4 & 0 & 0\end{pmatrix}\begin{pmatrix}x_1\\ x_2\\ x_3\end{pmatrix} = \begin{pmatrix}-2x_1\\ -7x_1 - 2x_2 - x_3\\ 4x_1\end{pmatrix} = \begin{pmatrix}0\\ 0\\ 0\end{pmatrix}$$

The solutions are as follows:

$$\begin{cases}x_1 = 0\\ x_3 = -2x_2\end{cases} \text{ and the eigenvector } \begin{pmatrix}x_1\\ x_2\\ x_3\end{pmatrix} = \begin{pmatrix}0\\ c_1\\ -2c_1\end{pmatrix} = c_1\begin{pmatrix}0\\ 1\\ -2\end{pmatrix}$$

where c_1 is a real nonzero number.

B. The eigenvectors corresponding to $\lambda = 1$

We get

$$\begin{pmatrix} 1-1 & 0 & 0 \\ -7 & 1-1 & -1 \\ 4 & 0 & 3-1 \end{pmatrix}\begin{pmatrix} x_1 \\ x_2 \\ x_3 \end{pmatrix} = \begin{pmatrix} 0 & 0 & 0 \\ -7 & 0 & -1 \\ 4 & 0 & 2 \end{pmatrix}\begin{pmatrix} x_1 \\ x_2 \\ x_3 \end{pmatrix} = \begin{pmatrix} 0 \\ -7x_1 - x_3 \\ 4x_1 + 2x_3 \end{pmatrix} = \begin{pmatrix} 0 \\ 0 \\ 0 \end{pmatrix}$$

and see that $\begin{cases} x_3 = -7x_1 \\ x_3 = -2x_1 \end{cases}$

But this could only be true if $x_1 = x_3 = 0$. So the eigenvector has to be

$$\begin{pmatrix} x_1 \\ x_2 \\ x_3 \end{pmatrix} = \begin{pmatrix} 0 \\ c_2 \\ 0 \end{pmatrix} = c_2\begin{pmatrix} 0 \\ 1 \\ 0 \end{pmatrix}$$

where c_2 is an arbitrary real nonzero number.

3. Since there were no eigenvectors in the form

$$c_2\begin{pmatrix} x_{12} \\ x_{22} \\ x_{32} \end{pmatrix} + c_3\begin{pmatrix} x_{13} \\ x_{23} \\ x_{33} \end{pmatrix}$$

for $\lambda = 1$, there are not enough linearly independent eigenvectors to cxpress

$$\begin{pmatrix} 1 & 0 & 0 \\ -7 & 1 & -1 \\ 4 & 0 & 3 \end{pmatrix} \text{ in the form } \begin{pmatrix} x_{11} & x_{12} & x_{13} \\ x_{21} & x_{22} & x_{23} \\ x_{31} & x_{32} & x_{33} \end{pmatrix}\begin{pmatrix} \lambda_1 & 0 & 0 \\ 0 & \lambda_2 & 0 \\ 0 & 0 & \lambda_3 \end{pmatrix}\begin{pmatrix} x_{11} & x_{12} & x_{13} \\ x_{21} & x_{22} & x_{23} \\ x_{31} & x_{32} & x_{33} \end{pmatrix}^{-1}$$

It is important to note that all diagonalizable $n \times n$ matrices *always* have n linearly independent eigenvectors. In other words, there is always a basis in R^n consisting solely of eigenvectors, called an *eigenbasis*.

LOOKS LIKE I GOT HERE A BIT EARLY...
HEY THERE, BEEN WAITING LONG?
ポーン
REIJ—
KYAA!
ドキッ
THAT VOICE!
AWW, DON'T BE LIKE THAT!
STOP IT! LET ME GO!
WE JUST WANT TO GET TO KNOW YOU BETTER.
MISA!

I HAVE TO DO SOMETHING...
AND FAST!
THOSE JERKS...
!
ガタ
ガタ
TREMBLE
BUT...
WHAT IF IT HAPPENS ALL OVER AGAIN?
ON MY FIRST DATE WITH YUKI, MY GIRLFRIEND IN MIDDLE SCHOOL...
HEY, LET ME GO!
WE JUST WANT TO HANG OUT...
I ALREADY HAVE A DATE.
YUKI!

THOSE JERKS...
I HAVE TO DO SOMETHING!
YOU... STOP IT!
UH...WHO ARE YOU?
カッ
KICK
GUH?
ドサッ
CRASH
HAHA
SERIOUSLY? ONE KICK? WHAT A WUSS...
TIME TO GO.
LEMME GO!
YOU THERE!

STOP. CAN'T YOU SEE SHE DOESN'T WANT TO GO WITH YOU?
JEEZ, WHO IS IT NOW?
OH NO!
Y-YOU'RE...
THE LEGENDARY LEADER OF THE HANAMICHI KARATE CLUB.
"THE HANAMICHI HAMMER!"
IN THE FLESH.
UNNG...
HANAMICHI HAMMER?
MESSING WITH THAT GUY IS SUICIDE!
LET'S GET OUTTA HERE!
T-THANK YOU!
DON'T WORRY ABOUT IT...BUT I THINK YOUR BOYFRIEND NEEDS MEDICAL ATTENTION...
MOAN...

LISTEN... I'M GLAD YOU STOOD UP FOR ME, BUT...
IT JUST WASN'T ENOUGH.
I...
...DON'T THINK I CAN SEE YOU ANYMORE.
...
I'M SO SORRY... I COULDN'T DO ANYTHING...
NOT THIS TIME.

I'LL SHOW THEM—
I'M SO MUCH STRONGER NOW—
COME ON, SWEETIE...
YANK
HELP!
THIS TIME WILL BE DIFFERENT!
LET HER GO!

SMACK
WHA–?

COME ON, LET'S GO.
REIJI!
HEY! STOP RIGHT THERE.

JUST WHO DO YOU THINK YOU ARE?!
ガ
BAFF
!

STAY AWAY FROM HER, ALL OF YOU!

HAHA, LOOK! HE THINKS HE'S A HERO!
LET'S GET HIM!

RUN!
BE CAREFUL, REIJI!
OOPH
SMACK
YOU LITTLE...
THUMP
STOP IT!
PLEASE!
LET ME GO!
STUBBORN, HUH?
ENOUGH!
GRAB

ATTACKING MY LITTLE SISTER, ARE WE?
TETSUO!
I DON'T LIKE EXCESSIVE VIOLENCE...BUT IN ATTACKING MISA, YOU HAVE GIVEN ME LITTLE CHOICE...
CRACK
IT'S ICHINOSE!
THE HANAMICHI HAMMER!
MOMMY!
RUN!
SENSEI?
HE'S OUT COLD.

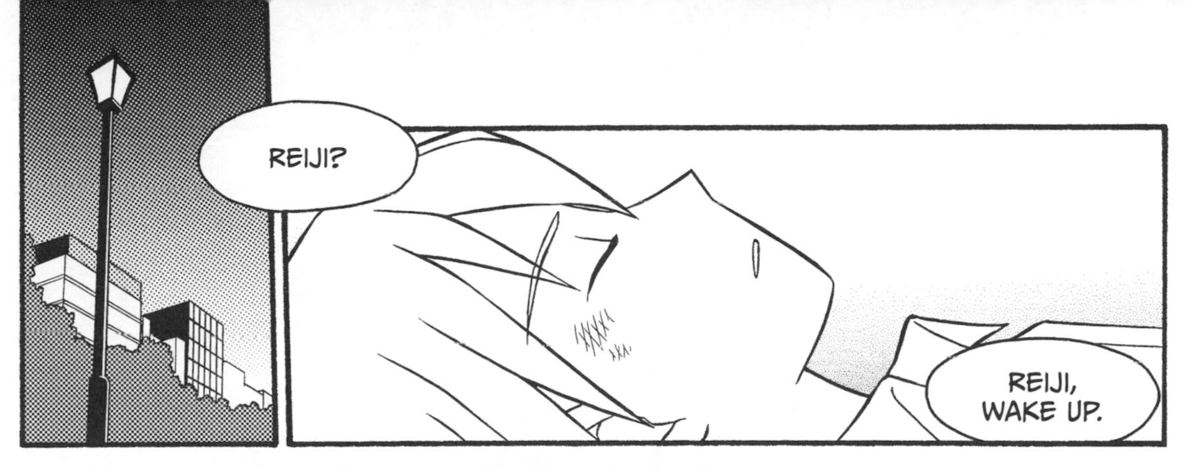
REIJI?
REIJI, WAKE UP.

REIJI!

YOU'RE OKAY!
WHOA!

BUT I DON'T DESERVE YOUR THANKS...

YURINO...MISA TOLD ME WHAT HAPPENED.
THANK YOU.
UM...NO PROBLEM.

I COULDN'T HELP MISA...I COULDN'T EVEN HELP MYSELF...
I HAVEN'T CHANGED AT ALL! I'M STILL A WEAKLING!

WELL, YOU MAY NOT BE A BLACK BELT YET...
BUT YOU'RE DEFINITELY NO WEAKLING.

PUTTING MISA'S SAFETY BEFORE YOUR OWN SHOWS GREAT COURAGE. THAT KIND OF COURAGE IS ADMIRABLE,
EVEN THOUGH THE FIGHT ITSELF WAS UNNECESSARY.

YOU SHOULD BE PROUD!
BUT—
REIJI!

HE'S RIGHT.
I DON'T KNOW WHAT TO SAY... THANK YOU.
MISA...

THANK YOU FOR EVERYTHING!
AH...
WHAT THE—!

I THOUGHT I WAS PRETTY CLEAR ABOUT THE RULES...
ギリ ギリ
HUH?! I, UH...

HEH...
WELL, I GUESS IT'S OKAY...MISA'S NOT A KID ANYMORE.
THANKS, SENSEI...

BY THE WAY, WOULD YOU CONSIDER DOING ME ANOTHER FAVOR?
S-SURE.

I'D LIKE YOU TO TEACH ME, TOO.
MATH, I MEAN.

WHAT?
HE COULD REALLY USE THE HELP, BEING IN HIS SIXTH YEAR AND ALL.
IF HE DOESN'T GRADUATE SOON...

SO. WHAT DO YOU SAY?
IT'D MEAN A LOT TO ME, TOO.
SURE! OF COURSE!
GREAT! LET'S START OFF WITH PLUS AND MINUS, THEN!
UM... PLUS AND MINUS?
SOUNDS LIKE YOU'LL NEED MORE LUNCHES!

ONLINE RESOURCES

THE APPENDIXES

The appendixes for *The Manga Guide to Linear Algebra* can be found online at *http://www.nostarch.com/linearalgebra*. They include:

Appendix A: Workbook
Appendix B: Vector Spaces
Appendix C: Dot Product
Appendix D: Cross Product
Appendix E: Useful Properties of Determinants

UPDATES

Visit *http://www.nostarch.com/linearalgebra* for updates, errata, and other information.

I STILL DON'T GET IT!
SNAP!
NO NEED TO GET VIOLENT...
YOU CAN DO IT, BRO!

INDEX

NOTES

NOTES

NOTES

ABOUT THE AUTHOR

Shin Takahashi was born 1972 in Niigata. He received a master's degree from Kyushu Institute of Design (known as Kyushu University today). Having previously worked both as an analyst and as a seminar leader, he is now an author specializing in technical literature.

Homepage: *http://www.takahashishin.jp/*

PRODUCTION TEAM FOR THE JAPANESE EDITION

SCENARIO: re_akino

ARTIST: Iroha Inoue

DTP: Emi Oda

HOW THIS BOOK WAS MADE

The *Manga Guide* series is a co-publication of No Starch Press and Ohmsha, Ltd. of Tokyo, Japan, one of Japan's oldest and most respected scientific and technical book publishers. Each title in the best-selling *Manga Guide* series is the product of the combined work of a manga illustrator, scenario writer, and expert scientist or mathematician. Once each title is translated into English, we rewrite and edit the translation as necessary and have an expert review each volume. The result is the English version you hold in your hands.

MORE MANGA GUIDES

Find more *Manga Guides* at your favorite bookstore, and learn more about the series at *http://www.nostarch.com/manga*.